DYNAMIC MASS SPECTROMETRY

DYNAMIC MASS SPECTROMETRY

Editors: D. Price (University of Salford)
J. E. Williams (FieldTech Ltd.)

Volume 1

The Second European Symposium on
the Time-of-Flight Mass Spectrometer

University of Salford, England, July 1969.

PUBLISHED BY **HEYDEN & SON LTD**
IN CO-OPERATION WITH **SADTLER RESEARCH LABORATORIES INC.**

Published in Great Britain by
Heyden & Son Ltd.,
Spectrum House, Alderton Crescent,
London NW4.

Co-published for exclusive distribution in the USA by
Sadtler Research Laboratories Inc.,
3314-20 Spring Garden St.,
Philadelphia, Pa. 19104.

© Heyden & Son Ltd., 1970.
All Rights Reserved. No part of this publication may be reproduced, stored in a retrieval system, or transmitted, in any form or by any means, electronic, mechanical, photocopying, recording or otherwise, without the prior permission of Heyden & Son Ltd.

Library of Congress Catalog Card No. 73-107612

SBN 85501 033 9

Printed in Great Britain by Avon Litho Ltd., Stratford-upon-Avon, Warwickshire, and bound at the Pitman Press, Bath, Somerset.

Contents

Preface   vii

KINETICS STUDIES

Chapter 1   The Vaporisation of Pyrolytic Graphite
P. D. Zavitsanos   1

Chapter 2   Kinetics Measurements of Fast Gas-Solid Reactions
V. Koch, K. H. van Heek and H. Jüntgen   15

Chapter 3   Multiple Pulse Radiolysis Studies of Inorganic Molecules with Low Energy Electrons
K. O. Dyson   23

Chapter 4   Flash Photodecomposition of Lead Tetramethyl Monitored by Time-Resolved Mass Spectrometry
S. E. Appleby, S. B. Howarth, A. T. Jones, J. A. Lippiatt, W. J. Orville-Thomas, D. Price and (in part) P. Heald   37

IONISATION PROCESSES

Chapter 5   An Application of Photoionisation in a Time-of-Flight Mass Spectrometer
P. F. Knewstubb and N. W. Reid   59

Chapter 6   Photoelectron Spectrometry and R.P.D. Measurements on Sulphur Hexafluoride
Jacques Delwiche   71

Chapter 7   The Use of an Ion Probe Technique for Investigating Surface Reactions: The Synthesis of Deutero-Ammonia on Pure Iron
J. C. Robb, D. R. Terrell and D. W. Thomas   87

Chapter 8   Studies of Negative Ion Formation at Low Electron Energies
P. W. Harland, K. A. G. MacNeil and J. C. J. Thynne   105

Chapter 9   Ion Spectroscopy by Nanosecond Resolution Time-of-Flight Techniques
            G. W. F. Pike   139

ANALYTICAL APPLICATIONS

Chapter 10 A Combined Field-Ion Microscope and Time-of-Flight Mass Spectrometer
            P. J. Turner and M. J. Southon   147

Chapter 11 The Use of Time-of-Flight Mass Spectrometry as a Selective Detector for Quantitative Gas Chromatography
            C. D'Oyly-Watkins, D. E. Hillman, D. E. Winsor and R. E. Ardrey   163

Chapter 12 The Identification of Cross-Linking Agents in Some Epoxy Resin Systems by Time-of-Flight Mass Spectrometry
            C. D'Oyly-Watkins and D. E. Winsor   175

Chapter 13 Laser Pyrolysis of Coal and Related Materials in the Source of a Time-of-Flight Mass Spectrometer
            W. K. Joy   183

GENERAL AND INSTRUMENTATION

Chapter 14 Resolution and Sensitivity of Mass Spectrometers
            D. C. Damoth   199

Chapter 15 A Linear-to-Logarithmic Compressor Circuit for Electronic Recorders
            K. O. Dyson   211

Chapter 16 An Amplitude Limiting Device for U.V. Galvanometer Recorders
            C. D'Oyly-Watkins and S. N. Gaythorpe   215

Chapter 17 A Gas Chromatograph/Mass Spectrometer Interface System
            A. J. Luchte and D. C. Damoth   219

Bibliography of Papers Involving the Use of Linear Non-Magnetic Time-of-Flight Mass Spectrometers
W. K. Joy   225

Index   243

Preface

Most of the chapters which form this book are papers presented at the 2nd European Time-of-Flight Mass Spectrometer Symposium and Users' School, held at the University of Salford, England, 7th-11th July, 1969. It is not a complete record of the proceedings of the Symposium, as some papers which were read have been omitted at their authors' request. One or two short papers have been added as they either supplement the work described in the main papers, or are of particular interest to time-of-flight mass spectrometer users. A report of the discussion following the presentation of each paper, has been included where it is of special interest.

The editors are indebted to Mr. W. K. Joy, of the British Coal Utilisation Research Association, for the preparation of the Bibliography.

Applications for fast-scanning dynamic mass spectrometers are far too numerous to be reported exhaustively in one book. This volume deals specifically with current work using the linear, non-magnetic time-of-flight instrument. It covers such diverse subjects as field-ion microscopy, negative ion formation, fast-reaction studies and gas chromatograph effluent monitoring. The time-of-flight mass spectrometer is shown to be a very versatile instrument due in the main to the fact that its construction lends itself to numerous forms of modification to meet specific experimental requirements. Like any other instrument, it has its limitations. It is hoped this volume will indicate the main areas of application to which it is best suited.

On behalf of the organising committee, the editors wish to thank the Bendix Corporation and FieldTech Ltd. for financial support and the University of Salford for the facilities used during the Symposium.

D.P., J.E.W.

Chapter 1

The Vaporisation of Pyrolytic Graphite

P. D. Zavitsanos

Re-entry and Environmental Systems
General Electric Company, King of Prussia, Pennsylvania, USA

Introduction

Pyrolytic graphite has been receiving a great amount of attention as a structural material in aerospace applications. The advantages are resistance to sublimation, high strength to density ratio at high temperatures, high oxidation resistance[1,2] (compared with other forms of graphite) and low thermal conductivity in the direction of the C-axis.

Since the vaporisation behaviour of high performance tips and heat shields constitutes an important input to aerodynamic calculations of weight loss, boundary layer chemistry and observables, it is significant to know: (1) gross rate of sublimation, (2) composition of the resulting vapour and partial pressures of the resulting species, (3) vaporisation coefficients for the gaseous carbon species so that the composition of the boundary layer can be predicted more accurately for non-equilibrium conditions. Incomplete knowledge of the above properties warranted the following study.

Experimental

Specimens. The specimens of pyrolytic graphite used in these studies were taken from a surface nucleated piece manufactured by the Metallurgical Products Department, General Electric Company, Detroit. The material had a density of 2·21 g cm^{-3} and high purity. The ash content was about 0·005% by weight. Spectroscopic analysis of the ash content is shown below:

Si	50%
Al	6
Ca	5
Zr	3
V	2
Fe	3
Ti	0·7
Ta	1

W	1
Ni	0.5
B	0.03
Oxygen	Bal.

Microbalance Target Technique. The vaporisation of pyrolytic graphite was studied by using the Knudsen effusion target method with a recording microbalance. A conical target, made of molybdenum sheet (or glass), intercepted a well-defined fraction of the vapour effusing from a Knudsen crucible. The weight increase of the target due to the condensation of carbon vapour was recorded by a Sartorius-Electrona microbalance. A detailed description of the apparatus with applications was given previously.[3,4] The crucibles used were made of tantalum or tungsten and had several orifice sizes. In general, the orifice sample surface ratio was always less than 10^{-3}. Electron bombardment was used for heating and the temperature was measured with a calibrated optical pyrometer by sighting into a black-body hole drilled on the side of the crucible. The optical pyrometer was calibrated against an NBS lamp. A quartz window used for sighting was kept clean (from vapours) with an iron slug which (actuated from outside with a small magnet) was removed from the window only during temperature measurements.

In addition to the effusion measurements, rates of sublimation were measured by heating cylinders of pyrolytic graphite in vacuum by electron bombardment. The cylinders (2.5 cm dia. x 1.5 cm long) were mounted in the same manner as the Knudsen crucible in the same apparatus. Temperature measurements were made pyrometrically by sighting into a black body hole drilled on the specimen. Rates of vaporisation from the plane perpendicular to the C-axis were obtained with the same microbalance target technique by measuring the condensation rate of carbon vapour vaporising from a surface area equal to the hole of the shield (1.1 cm²).

Mass Spectrometry. The vapour resulting from the vaporisation of pyrolytic graphite was analysed as a function of temperature under conditions of equilibrium and non-equilibrium using a Bendix time-of-flight mass spectrometer.

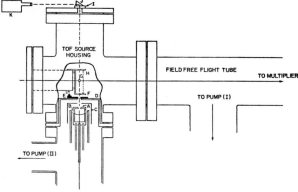

Fig. 1: Electron bombardment furnace and Bendix time-of-flight mass spectrometer.
A : Crucible; B : Tungsten Filament; C : Tantalum Shield; D : Shutter-Chamber; E : Movable Shutter; F : T.O.F. Source Entrance Slit; G : Ionising Electron Beam; H : Ion Grid; I : Viewing Window; J : Prism; K : Optical Pyrometer.

A diagram of the electron bombardment furnace and part of the mass spectrometer is shown in Fig. 1. The Knudsen crucible A, kept at positive voltage (0-5 kV), is heated by electron bombardment from a heated tungsten filament B. Temperatures as high as 2500°C were reached with control of ± 2°. Temperature control was achieved by keeping the emission current constant through automatic regulation of the filament current. A circuit diagram of the emission regulator was given elsewhere.[3] In order to prevent furnace ions from entering the mass spectrometer and also to reduce background noise, negative voltage was applied on shield C. By keeping the absolute value of the biasing negative voltage slightly higher than the positive voltage of the crucible, ions with positive charge are either collected or deflected by the shield so that they do not enter the ion source of the mass spectrometer. A grounded crucible-negative filament arrangement was also used. This second heating arrangement appeared to give less background noise and is considered preferable. The temperature was measured by sighting into the orifice and with the additional correction for the prism J (Fig. 1).

After proper collimation, a small fraction of the vapour effusing from the cell enters the ionisation region of the ion source where a pulsed electron beam (directed perpendicular to the neutral beam) ionises a fraction of the gas; mass analysis is carried out by the time-of-flight method. Non-equilibrium (Langmuir) vaporisation was also carried out in the mass spectrometer. This was done by replacing the crucible with a cylindrical specimen of pyrolytic graphite. The plane perpendicular to the C-axis was facing the ionisation region of the mass spectrometer and temperature measurements were made by sighting at the surface. Emissivity measurements by DeSantis[5] were used to correct the pyrometer readings. The emissivity corrections were checked in a separate experiment by looking at a black body hole and the basal plane with the optical pyrometer.

Theory

Knudsen Effusion. Knudsen crucibles with three different orifice-sizes (0·0109, 0·0126, 0·0248 cm^2) were used in the microbalance experiments. A collimator with a radius $r = 0.3$ cm placed at a distance $c = 1.5$ cm from the crucible allows a fraction of the effusing vapour to strike the target. Increases in the weight of the target (ΔW) as a function of time (Δt) are shown in Table 1. The condensation rate m' (gm sec^{-1}) is

$$\frac{10^{-6} \Delta W}{60 \Delta t}$$

Effusion rates m are then obtained from

$$m = m' \left[(c^2 + r^2)/r^2 \right] A^{-1} \quad (1)$$

where A is the area of the orifice. Rates of effusion obtained in the temperature range 2417-2700° K are shown in Table 1. Comparison with the effusion data of Brewer *et al.*[6] taken on spectroscopic grade graphite shows good agreement. The equilibrium vapour pressure, P, is to be calculated from the Knudsen equation

$$P = m (2\pi RT/M)^{1/2} \quad (2)$$

Table 1. Vaporisation of pyrolytic graphite, Knudsen microbalance technique

Run No.	T (°K)	ΔW (μg)	Δt (min)	$\dfrac{c^2 + r^2}{r^2}$	$A \times 10^2 \text{cm}^2$	Effusion Rate $m \times 10^7$ gm cm^{-2} sec^{-1}
19-A-1	2564	12	61	22·436	1·095	36·6
19-A-3	2595	34	178	22·436	1·266	56·41
19-A-4	2562	29	180	14·58	1·095	27·873
19-A-5	2518	26	320	14·580	1·266	17·702
20-A-2	2565	38	330	14·580	1·095	25·518
56	2422	10	235	10·269	2·48	2·936
57	2452	25·5	180	10·269	2·48	9·7931
59	2417	25	240	10·269	2·48	7·1716
61	2515	29	120	10·269	2·48	16·62
66	2482	24	330	10·269	2·48	4·231
67	2700	251·8	100	10·269	2·48	173·8

where m is the effusion rate in gm cm^{-2} sec^{-1}, T is the temperature in °K, and M is the molecular weight of the vapour. In the case of graphite, the vapour consists of atomic as well as molecular species (i.e. C_1, C_2, C_3, C_4, C_5 .....) and in these experiments the total effusion rate is measured, $m = \sum_i m_i$ As shown by previous mass spectrometric work [7,8] and verified by this investigation, the important species, in the temperature range of this work, are C_1, C_2, and C_3 and therefore

$$m = \left(\frac{12}{2\pi RT}\right)^{1/2} \left\{ P_{C_1} + \sqrt{2}\, P_{C_2} + \sqrt{3}\, P_{C_3} \right\} \quad (3)$$

By measuring ratios of ion intensities I_n/I_l, for the species in question by the time-of-flight mass spectrometer, the following expressions can be used:

$$\frac{P_{C_2}}{P_{C_1}} = \frac{I_2}{I_1} \times \frac{\sigma_1}{\sigma_2} \times \frac{(E - A_1)}{(E - A_2)} \quad (4)$$

$$\frac{P_{C_3}}{P_{C_1}} = \frac{I_3}{I_1} \times \frac{\sigma_1}{\sigma_3} \times \frac{(E - A_1)}{(E - A_3)} \quad (5)$$

where σ = relative cross section for ionisation, E = energy of ionising electrons, A = appearance potential ($A_1 = 11\cdot3, A_2 = 12, A_3 = 12\cdot6$ eV). All the results were obtained at a low electron energy (i.e. $E = 20$ eV) in order to avoid fragmentation of the molecular species. Additivity of cross sections were assumed[9] i.e.

$$\sigma_1/\sigma_2 = 1/2; \; \sigma_1/\sigma_3 = 1/3$$

Mass spectrometric data obtained in the temperature range 2500-2755° K are shown in Table 2.

Table 2. Ion intensity ratios $\dfrac{I_n}{I_1}$ from the Knudsen mass spectrometric techniques for C_1, C_2, C_3

T (°K)	$\dfrac{I_2}{I_1}$	$\dfrac{I_3}{I_1}$
2500	0·30	5·00
2564	0·35	5·25
2700	0·35	5·30
2703	0·40	5·33
2708	0·50	6·00
2722	0·39	4·67
2728	0·4	4·20
2745	0·43	6·22
2755	0·46	4·64
2755	0·71	6·64

Vapour pressures obtained from the combination of microbalance and mass spectrometric techniques through Eqns. (3), (4) and (5) are shown in Table 3. Using these vapour pressure data and free energy functions from the JANAF Tables,[10] heats of sublimation were calculated for C(g), C_2(g) and C_3(g) by the third law method:

$$\Delta H^\circ_{298} = T \left\{ -R\ln P + \left(\dfrac{G^\circ_T - H^\circ_{298}}{T}\right)_{solid} - \left(\dfrac{G^\circ_T - H^\circ_{298}}{T}\right)_{gas} \right\} \quad (6)$$

These values, 173·5, 203·5 and 192·0 kcal mole^{-1} respectively, also shown in Table 3, are to be compared with the best literature values listed in Table 4.

In view, however, of more recent spectroscopic work on the C_3 molecule, its heat of sublimation deserves additional attention: Gausset et al.[15] have shown that the bending frequency of C_3 is around 63 - 70 cm^{-1} instead of 550 cm^{-1} on which the JANAF thermodynamic functions are based.

Free energy functions based on frequencies 63 - 70 (degenerate), 1235, and 2040 cm^{-1} made by Weltner[16] — who assumed an harmonic oscillator rigid rotator — and Robiette and Strauss[17] — who took account of the considerable quartic anharmonicity of the bending vibration — are shown in Table 5.

Values of $\dfrac{G^\circ - H^\circ_0}{T}$ were converted to $\dfrac{G^\circ - H^\circ_{298}}{T}$ by adding

Table 3. Vapour pressure and calculated values of ΔH°_{298} of vaporisation for C(g), C_2(g), C_3(g).

	Log P_{C_n} (atm)			ΔH°_{298} (kcal mole⁻¹) for: $nC(s) \rightarrow C_n(g)$		
T (°K)	$n = 1$	$n = 2$	$n = 3$	$n = 1$	$n = 2$	$n = 3$
2564	−6·572	−7·312	−6·284	173·17	203·38	191·64
2595	−6·393	−7·132	−6·104	173·14	203·65	191·72
2562	−6·702	−7·442	−6·414	174·56	204·73	193·02
2518	−6·903	−7·642	−6·614	173·89	203·56	192·11
2565	−6·740	−7·480	−6·451	175·22	205·43	193·68
2422	−7·692	−8·431	−7·403	176·00	204·64	193·71
2452	−7·166	−7·905	−6·877	172·29	201·28	190·02
2417	−7·301	−8·040	−7·022	171·34	199·95	189·09
2515	−6·931	−7·670	−6·642	174·01	203·67	192·19
2482	−7·137	−7·876	−6·848	174·06	203·34	192·00
2700	−5·900	−6·640	−5·611	174·01	205·57	193·18
			Average.	173·56±1,	203·5±1·1,	192·01±1·1

Table 4. Heats of sublimation of carbon species

Species	ΔH°_{298} kcal mole⁻¹	Reference
C_1	171·29 (±0·5)	Recommended by 11
	173·5 (± 1)	This Work
C_2	197·02	12
	199·02	13
	200 (± 1·7) (by 3rd law)	7
	197·8 (± 1·7) (by 2nd law)	7
	203·5 (± 1)	This Work
C_3	200·5	13
	190·5	14
	189·7 (± 2·3) (by 3rd law)	7
	186·7 (± 1·5) (by 2nd law)	7
	192·0 (± 1·1)	This Work

Table 5. Thermodynamic functions for C_3 using a low bending frequency (70 - 63 cm^{-1})

T (°K)	$S°_{16}$	$\dfrac{(G°-H°_{298})_{16}}{T}$	$S°_{17}'$	$\dfrac{(G°-H°_{298})_{17}}{T}$
2400	84·993	73·073	82·796	71·25
2500	85·589	73·562	83·356	71·72
2600	86·162	74·036	83·895	72·18
2700	86·715	74·495	84·415	72·63
2800	87·248	74·941	84·916	73·05
3000	88·261	75·796	85·868	73·88
3200	89·210	76·605	86·760	74·66
3400	90·103	77·373	87·599	75·39
3600	90·946	78·104	88·391	76·09
3800	91·744	78·801	89·141	76·76
4000	92·502	79·467	89·854	77·40
4200	93·224	80·105	90·531	78·01
4400	93·912	80·717	91·78	78·59
4600	94·571	81·305	91·796	79·15

$\dfrac{H°_0 - H°_{298}}{T}$, $H°_0 - H°_{298} = 2·32$ kcal.

If these new values for the free energy functions are used in the third law calculation considerably higher values are obtained for the heat of vaporisation:

$$\Delta H°_{298} = 212 \text{ kcal mole}^{-1} \text{ (based on reference 16)}$$

$$\Delta H°_{298} = 207 \text{ kcal mole}^{-1} \text{ (based on reference 17)}$$

These values depart significantly from the second law value of 186·7 ± 1·5 as reported by Drowart et al.[7] Using the higher heats and the free energy functions of Table 5, vapour pressures were calculated in the temperature range 2800 - 4600°K from Eqn. (6). Fig. 2 shows calculated vapour pressures based on $\Delta H°_{298} = 207$ kcal mole^{-1} and free energy functions as calculated by Strauss.[17]

Langmuir Vaporisation. Since most of the vaporisation phenomena of interest involve non-equilibrium conditions, it is of interest to know the vaporisation coefficients for the various carbon species so that respective rates of vaporisation (m) can be calculated:

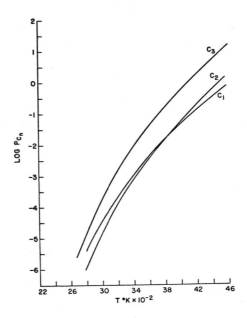

Fig. 2: Calculated vapour pressure for C, C_2 and C_3

$$m_i = \alpha_i P_i \left(\frac{M_i}{2\pi RT} \right)^{1/2} \tag{7}$$

where $i = 1, 2, 3$
α = vaporisation coefficient
P = equilibrium vapour pressure
M = molecular weight
T = temperature in °K

By carrying out weight loss measurements in vacuum as stated before, the total rate of vaporisation m was determined. This quantity can be expressed as

$$m = \sum_{i=1}^{3} \alpha_i P_{C_i} \left(\frac{M_i}{2\pi RT} \right)^{1/2} \tag{8}$$

and is shown in Table 6.

By measuring relative ion intensities for the three predominant species (C, C_2 and C_3), in the mass spectrometer under both equilibrium and free vaporisation conditions, I and I' respectively, ratios of vaporisation coefficients can be calculated independently of any knowledge of relative cross sections and multiplier efficiencies:

For equilibrium conditions:

$$\frac{P_{C_2}}{P_{C_1}} = \frac{I_2}{I_1} \times \frac{\alpha_1}{\alpha_2} \times \frac{(E - A_1)}{(E - A_2)} \tag{9}$$

Table 6. Vaporisation of pyrolytic graphite (from the plane perpendicular to the C-axis); Langmuir microbalance technique

Run No.	T (°K)	Rate $m \times 10^7$ (gm/cm² sec)
70	2460	0·80
71	2273	0·02
72	2305	0·100
73	2316	0·037
73-A	2460	0·757
78	2400	0·317
79	2540	2·145
80	2300	0·108
82	2250	0·045
83	2250	0·056
86[a]	2560	2·241
87[a]	2500	1·394
88[a]	2560	2·094
89[a]	2560	2·902

[a] Mechanically polished surface

and for non-equilibrium vaporisation

$$\frac{\alpha_2 P_{C_2}}{\alpha_1 P_{C_1}} = \frac{I'_2}{I'_1} \times \frac{\alpha_1}{\alpha_2} \times \frac{(E-A_1)}{(E-A_2)} \tag{10}$$

Combining Eqns. (9) and (10)

$$\frac{\alpha_2}{\alpha_1} = \frac{I'_2}{I'_1} \times \frac{I_1}{I_2} \tag{11}$$

Similarly for

$$\frac{\alpha_1}{\alpha_3} = \frac{I'_1}{I'_3} \times \frac{I_{-3}}{I_{-1}} \tag{11a}$$

Use of Eqns. (11), (11a), and (8) can be made to calculate vaporisation coefficients for C, C_2 and C_3.

Relative ion intensities measured in the temperature range 2618 - 2760°K for the free vaporisation from the plane perpendicular to the C-axis are shown in Table 7.

Combination with data from Tables 2, 3 and 7 gives values for the vaporisation coefficients, $\alpha_1 = 0.24$, $\alpha_2 = 0.5$, $\alpha_3 = 0.023$, also shown in Table 7.

Table 7. Free vaporisation of pyrolytic graphite (from the plane perpendicular to the C-axis).

T (°K)	$\dfrac{I'_2}{I'_1}$	$\dfrac{I'_3}{I'_1}$	$\dfrac{\alpha_2}{\alpha_1}$	$\dfrac{\alpha_1}{\alpha_3}$
2760	0·89	0·61	1·95	9·24 ± 1·6
2740	0·91	0·56	2·1	11·11
2700	0·86	0·52	2·09 ± 0·25	10·6 ± 0·6
2660	0·80	0·48	2·28	10·93
2618	0·75	0·46	2·14	11·41
Average:			2·1	10·6

$$\alpha_1 = 0.24$$
$$\alpha_2 = 0.50$$
$$\alpha_3 = 0.023$$

The overall vaporisation coefficient $\left[\dfrac{\Sigma \alpha_i P_i \sqrt{i}}{\Sigma P_i \sqrt{i}} \right]$ suggested by these measurements is 0·095 which is to be compared with the values of 10^{-3} observed by Doerhard, Goldfinger, and Waelbroeck,[16] 0·15 reported by Thorn and Winslow,[14] and 0·07 most recently reported by Burns, Jason and Inghram.[17]

The significance of the very low vaporisation coefficient of C_3 is that although C_3 is the most predominant species under equilibrium conditions, this is no longer the case when equilibrium is not present. To demonstrate the significance of vaporisation coefficients, calculated vaporisation rates for C, C_2, and C_3 are shown in Fig. 3 up to 4000°K for (a) the case of measured α's (b) for the case where α's are assumed to be unity (an assumption usually made in the absence of experimental information). It would also be interesting to make use of the non-equilibrium behaviour of graphite vapour in the analysis of Scala and Vidale[20] and Scala and Gilbert[21] to determine the range where the vaporisation rates (under flight conditions) are no longer controlled by an equilibrium-diffusion process. Considering the actual α's for C, C_2 and C_3, Scala and Gilbert[22] have obtained non-equilibrium heterogeneous criteria for each species vaporising from graphite for one hypersonic trajectory. The results indicate that the vaporisation rates for C and C_2 are controlled by the equilibrium-diffusion process while the rate for C_3 is controlled by non-equilibrium vaporisation.

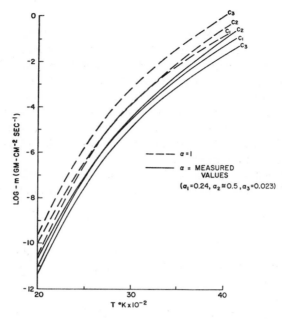

Fig. 3: The effect of vaporisation coefficients on the composition of carbon vapour

Discussion and conclusions

This work is a rather complete study on a historically controversial subject. The attempt here was to use several complementary techniques by the same investigator and to study the vaporisation properties of pyrolytic graphite (which is a rather pure form of graphite). The microbalance target technique used in this study has advantages and disadvantages as previously discussed.[4] In the case of graphite vaporisation, the main problem is the uncertainty in the sticking coefficient of carbon vapour at the target temperature. The high reflection coefficient for C_3 measured by Chupka *et al.*,[23] it is believed, could not have affected these microbalance results significantly, since the target was kept at much lower temperatures. Further the conical shape of the target forces the vapour species to collide with the surface more than once, if the first collision does not result in condensation.

Effusion-rate and mass-spectrometric data were used to obtain vapour pressures for C, C_2 and C_3 which are essentially in agreement within less than a factor of two with the vapour pressures reported by Drowart *et al.*[7] based on mass spectrometric work. The heat of vaporisation of C_3 is in some doubt in view of the conflict between the second and third-law values.

Since the same (microbalance) technique was used for the free vaporisation as in the effusion studies, it is expected that reflection effects are even less important in the case of vaporisation coefficients; this effect would cancel out. The effect of type of graphite, surface characteristics and crystallographic plane on vaporisation coefficients could very well account for the small difference in α's between this work and the work of Burns *et al.*[19] and Thorn and Winslow.[14] The very low value of 10^{-3} reported by Doerhard *et al.*[16] is obviously in error.

Acknowledgements

This work was sponsored by the U. S. Air Force, Ballistic Systems Division, under Contract AF 04(694)-222 and released 10 October 1964 for public distribution as an unclassified document. Permission to publish is gratefully acknowledged.

The author is grateful to R. G. Brownlee who obtained the experimental data, to Prof. Leo Brewer of the University of California for constructive comments and for bringing to the author's attention the unpublished work of Robiette and Strauss.
Dr. H. L. Strauss of the University of California and Dr. W. Weltner Jr. of the University of Florida furnished the author with unpublished data on the free energy functions of C_3 and their kindness is much appreciated.

Dr. H. L. Friedman and G. Griffith made available the electronics of their mass spectrometer to the author; their help in obtaining the data and their constructive comments are also appreciated.

References

1. W. S. Horton, ATL – G.E. Co. Report No. 60GL218, January 1961.
2. P. D. Zavitsanos, Semi-Annual Report for Advanced Re-Entry Programme. Contract No. AF 04(694) - 667, June 1964.
3. P. D. Zavitsanos, *Rev. Sci. Instr.* **35**, 1061 (1964).
4. P. D. Zavitsanos, *J. Phys. Chem.* **68**, 2899 (1964).
5. V. DeSantis, et al., *Carbonisation of Plastics and Refractory Materials,* Wright Field Report, (Contract No. AF 33(616) - 6841) March 1965.
6. L. Brewer, P. W. Gilles and F. A. Jenkins, *J. Chem. Phys.* **16**, 797 (1948).
7. J. Drowart, R. P. Burns, G. DeMaria and M. G. Inghram, *J. Chem. Phys.* **31**, 1131 (1959).
8. W. A. Chupka and M. G. Inghram, *J. Chem. Phys.* **21**, 371 (1953): **21**, 1313 (1959).
9. J. W. Otros and D. P. Stevenson, *J. Chem. Phys.* **78**, 546 (1956).
10. JANAF, *Thermochemical Tables,* Dow Chemical Company, Midland, Michigan, 1961.
11. W. H. Evans, *Nat. Bur. Standards Rept.* 8504 (1964).
12. L. Brewer, W. J. Hicks and O. H. Krikorian, *J. Chem. Phys.* **36**, 182 (1962).
13. W. A. Chupka, and M. J. Inghram, *Phys. Chem.* **59**, 100 (1955).
14. R. J. Thorn, and G. H. Winslow, *J. Chem. Phys.* **26**, 186 (1957).
15. L. Gausset, G. Herzberg, A. Lagerquist, and B. Rosen, *Discussions Faraday Soc.* **35**, 113 (1963); *Astrophys. J.* **142**, 45, (1965).
16. W. Weltner, Jr. and D. McLeod, Jr. *J. Chem. Phys.* **40**, 1305 (1964); also private communication, 1966.
17. A. G. Robiette, and H. L. Strauss, *J. Chem. Phys.* **44**, 2826 (1966), also H. L. Strauss, *J. Chem. Phys.* **45** (1966).
18. Doerhard, Goldfinger and Waelbroeck, *Bull. Soc. Chem. Belges* **62**, 498 (1953).
19. R. P. Burns, A. J. Jason and M. G. Inghram, *J. Chem. Phys.* **40**, 1161 (1964).
20. S. M. Scala and G. Vidale, *Intern. J. Heat Mass Transfer* **1**, 4 (1960).
21. S. M. Scala, and L. Gilbert, *AIAA Journal* **3**, 1635 (1965).
22. Private communication.
23. W. A. Chupka, J. Bekowitz, D. J. Meschi, and H. A. Tasman, *Advances in Mass Spectrometry,* Vol. 2, Pergamon Press, Oxford, 1963, p. 9.

Discussion

Dr. van Heek: How is the 'laser induced' temperature on a graphite surface measured or estimated?

Dr. Zavitsanos: The temperature was not measured; it was calculated, based on JANAF Tables, and the measured ratios of C_2/C_4 and C_2/C_5 which depend very strongly on temperature (for more details see *Carbon* **6**, 731 (1968)).

Chapter 2

Kinetics Measurements of Fast Gas-Solid Reactions

V. Koch, K. H. van Heek and H. Jüntgen

*Bergbau-Forschung GmbH, Forschungsinstitut des Steinkohlenbergbauvereins,
Essen-Kray, W. Germany*

Introduction

Thermal reactions between gas and solids play a great part in technology and industry. It may be sufficient to mention the conversion of carbonates to their oxides, e.g. in limestone and cement works, the carbonisation or gasification of coal and also the combustion or explosion of solid fuels. Frequently, these reactions proceed very quickly at high temperatures (i.e. in less than one second) which is due partly to the development of modern technology and partly to the nature of the process. Thus, it seems necessary to investigate the fundamentals of gas-solid reactions under these conditions in the laboratory. The time-of-flight mass spectrometer is particularly appropriate to determine the products of fast decomposition reaction. This instrument has been used by several investigators[1-3] to study the formation of species from coal or graphite heated by a laser beam. This chapter describes a technique which has been developed to investigate the kinetics of fast gas-solid reactions. This has been used to establish correlations between the progress of the reaction, time and the temperature of the samples.[4] The experimental problems involved in the investigations will first be discussed and then some results of the experiments will be presented.

Heating of the samples and determination of the temperature

Figure 1 shows the heating system which was used for vacuum pyrolysis of solid samples on gauze wire cloth.[5] The width of the meshes of the gauze is such that the solid particles are surrounded from all sides by the heating wires; this ensures good heat contact. The gauze is heated up linearly with the time by an electric current; heating rates of several thousand °C/sec can be reached. On account of the small quantity of the sample it is impossible to measure directly the temperature of the solid particles. A fine thermocouple — the diameter of the legs is only 50 microns — whose junction is also embedded into the meshes of the gauze is used. In addition, the surface temperature is measured by an automatic radiation pyrometer whose time of response is 10 msec.

A disadvantage of thermocouples is the delay of their indications in the case of high rates of heating. The pyrometer records the values almost without delay but owing to the gases formed by the reaction, the values are disturbed at random. A combination of the

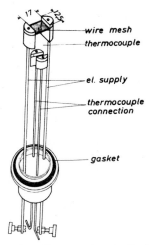

Fig. 1: Solid materials can be pyrolysed on an electrically heated wire gauze cloth of stainless steel

two measuring instruments, however, permits compensation of their specific errors.[6]

The sample of solids was spread on the gauze over a surface area of a few square millimetres. The next stage of the investigation was to check whether uniform heating of the sample was obtained. Measurements of the temperature distribution over the wire gauze cloth showed that the maximum temperature difference over an area of 35 mm² was 30°C. As the surface of the sample was much smaller, it can be assumed that the temperature difference does not exceed 5 to 10°C. It can be derived from theoretical investigations[4, 6] that at a rate of heating of 10^4 °C/min the ideal course of a reaction is not substantially affected by a temperature difference of 10°C.

Transfer of the gases into the ion source of the T.O.F. mass spectrometer

The heating device represented in Fig. 1 is accommodated in a high-vacuum

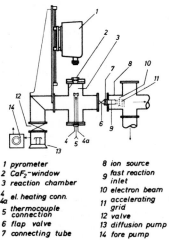

1 pyrometer
2 CaF$_2$-window
3 reaction chamber
4, 4a el. heating conn.
5 thermocouple connection
6 flap valve
7 connecting tube
8 ion source
9 fast reaction inlet
10 electron beam
11 accelerating grid
12 valve
13 diffusion pump
14 fore pump

Fig. 2: The decomposition reactions are performed in a vacuum chamber (10^{-4} torr) which is connected to the T.O.F. mass spectrometer (Bendix 14-107)

chamber from which the gas is withdrawn by a high capacity diffusion pump. A plate valve connects the chamber to the ion source of the T.O.F. mass spectrometer (Fig. 2).

The gas formed during the heating of the sample must be pumped off as quickly as possible so that the decisive factor for the readings of the spectrometer is the formation of the gas and not the pressure increase in the reaction chamber. For the reactions in question a time of response of some hundredths of a second is required. This corresponds to the dead time of the analogue amplifier which was used for the measurements. This presupposes an appropriate correlation between the pumps and the conductance of the reaction chamber and the connecting pipes.[7]

Figure 3 shows a diagrammatic flowsheet and indicates the volumes, the conductance values, and the effective pumping rates of the inlet system used in our experiments and represented in Fig. 2.

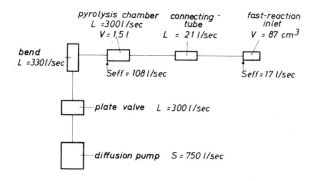

Fig. 3: The vacuum pyrolysis device, whose flowsheet is shown, has a time of response of less than 0·1 sec for the transfer of gases from the pyrolysis chamber into the ion source.

With the values quoted, the effective pumping rate in the reaction chamber is 108 litres/sec which corresponds to a pumping time of 0·05 sec for a pressure drop from 10^{-4} to 5×10^{-6} torr. This dead time can be easily checked. Air is admitted through a leak into the reaction chamber so as to produce a pressure of 10^{-4} torr. At the same time the nitrogen peak for the mass 28 is measured. If the valve is shut suddenly, the corresponding signal drops within 0·07 sec from 90% of its full height to 10%. This means that the real pumping rate in the reaction chamber is 50 litres/sec. That is to say the time of response of the apparatus with full pumping speed is at any rate less than one tenth of a second.

Results of the fast vacuum pyrolysis

Figure 4 shows the progress of carbon dioxide formation when calcium carbonate is decomposed at a rate of heating of $9 \times 10^3 \,°C/min$. A normalised reaction rate is plotted against the temperature. The full line represents the measured curve which leads to an activation energy of 45 kcal/mole and a frequency factor of 1×10^{11} min^{-1}, using the evaluation methods of non-isothermal reaction kinetics.[8] The curve which was calculated from these values, assuming a first order reaction, is drawn in dotted lines; the calculated

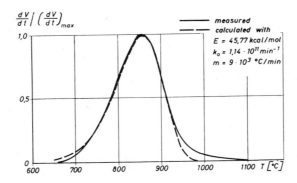

Fig. 4: The calcination of calcium carbonate can be described as a reaction of first order with the given values for the activation energy E and the frequency factor k_0

and the experimentally established curves agree very well.

The data shows that the calcination is governed by a first order law which was also found for lower rates of heating, e.g. in a differential sweep gas reactor. The apparent reaction parameters which were found in this case (E and k_0) are somewhat lower than those measured under normal pressure, which is due to the fact that the decomposition of carbonates is a reversible reaction and, therefore, the partial pressure plays a part.[9]

Another example is the release of ethane during the pyrolysis of different coals at rates of heating between 10^4 and 10^5 °C/min. This process is also governed by a first-order kinetic law. The results can be compared directly with those measured at very much smaller heating rates, down to 10^{-2} °C/min carried out in different experimental devices,[10] taking into account that the pyrolysis reaction is an irreversible process whose kinetics are in a wide range independent of pressure. The simplest way to do this is to use the theoretically expected shift of the temperature of the highest gas emission rate (T_M) towards higher temperature ranges with increasing rates of heating (m).[11]

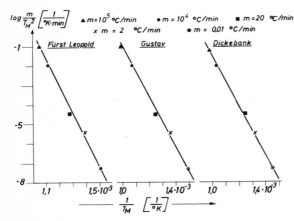

Coal	Volatile matter (m.a.f.) (%)
Fürst Leopold	39
Gustav	29
Dickebank	19

Fig. 5: A special kinetic plot of the ethane formation from coals, taking into account the shift of T_M (temperature of the maximum gas emission rate) due to rising rates of heating m, shows a good agreement of the measurements with the theory

Figure 5 is a plot of the measured values of T_M for given value of m in a log (m/T_M^2) versus $1/T$ diagram. The points which were measured at high rates of heating, 10^5 and 10^4°C/min, are well in line with those found at heating rates of 20·2 and 0·01°C/min, with only small deviations from the straight line which theoretically is to be expected. The essential conclusion of the experiments is that the kinetics of gas formation do not change over a wide range of heating rates as applied in our tests, and that the experimental conditions of the technique established are well defined. Moreover, the experiments are likely to facilitate the transfer of the results of laboratory tests obtained at low rates of heating to technical processes which are carried out at high rates of heating. [12,13]

Reactions under normal pressure

For reactions under normal pressure the heating system is accommodated in a glass vessel (Fig. 6) which can be filled with and traversed by different gases. Above the wire gauze there is a heated capillary tube through which the gases can be directed to the inlet system. At the end of the capillary tube the gas is first discharged into a vacuum

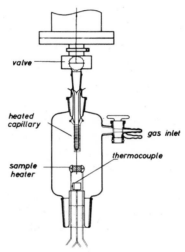

Fig. 6: The fast combustion of single coal grains is performed in an air flown glass vessel

chamber of 10^{-4} torr, the same that was used as reaction chamber in the high-vacuum test previously described.

As an example of the gas/solid reactions which can be carried out in this reactor the combustion of individual coal grains in air was examined. The original curve in Fig. 7 depicts the course of the carbon dioxide formation during the combustion of the coal grain when the wire gauze is heated at a rate of 10^4°C/min from room temperature to 1200°C. Two clearly separated peaks can be distinguished on the graph.

This pattern reflects the combination of pyrolysis and combustion when the grain of coal is heated up in air. The first step during the heating of the grain is the release of volatile liquid and gaseous hydrocarbons; together with the oxygen of the air they form around the grain a highly explosive layer which is ignited and burnt. This is the cause of the first very sharp peak of carbon dioxide formation. The remainder of the grain is then consumed at a much lower rate of carbon dioxide emission.[14]

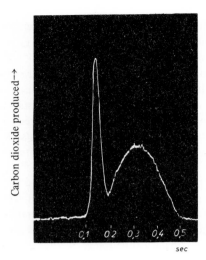

Fig. 7: During the combustion of coal under heating rates $10^4\,°C/min$ carbon dioxide is produced in two different steps, due to the burning of the volatile hydrocarbons and the solid material

It should be mentioned, however, that the picture of the co-action of pyrolysis and combustion observed at a heating rate of $10^4\,°C/min$ will not necessarily hold true if coal grains are heated up in air at higher rates of heating. If the heating rate is increased to say $10^6\,°C/min$, the solid coal grains can be ignited prior to gas emission.[15]

These experiments were performed to obtain a better understanding of the combustion of coal in boilers and conditions for dangerous coal dust explosions in mines.

Acknowledgement

The studies described in this paper have been sponsored by the European Community of Coal and Steel within the framework of the Fundamental Research Programme of 'Physics and Chemistry of Coal and Coke'.

References

1. W. K. Joy, W. R. Ladner and E. Pritchard, *Nature* **217**, 640 (1968).
2. F. J. Vastola and A. J. Pirone, *Prep. Div. Fuel Chem. Am. Chem. Soc.* **10** (2) C 53 (1966).
3. P. Zavitsanos, *Carbon* **6**, 731/37 (1968).
4. V. Koch, K. H. van Heek and H. Jüntgen, *Ber. Bunsenges. Phys. Chem.* **73**, to be published (1969).
5. R. Loison and R. Chauvin, *Chim. Ind. (Paris)* **94**, 269/75 (1965).
6. V. Koch, *Ph.D. Thesis,* Aachen, 1968.
7. P. A. Redhead, *Vacuum* **12**, 203/11 (1962).
8. K. H. van Heek and H. Jüngten, *Ber. Bunsenges. Phys. Chem.* **72**, 1223/31 (1968).
9. H. Jüntgen and K. H. van Heek, *Fortschr. Chem. Forsch.* to be published, 1969.
10. H. Jüntgen and K. H. van Heek, *Fuel* **47**, 103/17 (1968).
11. V. Koch, J. Jüntgen and W. Peters, *Brennstoff-Chem.* **50**, to be published (1969).
12. W. R. Ladner, *BCURA Bull.* **28**, 281/301 (1964).

13   W. Reerink and W. Peters, *Brennstoff-Chem.* **46**, 330/43 (1965).
14   V. Koch, H. Jüntgen and W. Peters, paper submitted to the conference *Verbrennung und Feuerung,* Karlsruhe, September 1969.
15   J. B. Howard and R. H. Essenhigh, *Ind. Eng. Chem. Des. Dev.* **6**, 74 (1967).

Discussion

Dr. Zavitsanos: The decomposition of calcuim carbonate has been studied by the isothermal technique as well. Would you please comment on the comparison between kinetic parameters obtained by that technique and yours.

Dr. van Heek: A comprehensive discussion of this point is given in my reference No. 9. An exact comparison is difficult, as the kinetic parameters depend firstly on the kind of the sample; crystalline structure, purity, grain size — and secondly on the experimental conditions; kind of heat and mass transfer, partial pressure of the carbon dioxide, technique and rate of heating. The values for the activation energy, frequency factor and reaction order estimated from the experiments are, therefore, apparent values in those cases in which the experimental conditions are not well defined or in which the chemical reaction does not determine the velocity of the process. For this reason, most of the values given in literature are not comparable, varying over a wide range both under isothermal and non-isothermal conditions. Values for the order of the reaction between 0 and 1 and activation energies from 35 to 50 kcal/mol are given by different authors. In our experiments, however, the experimental conditions are chosen in such a way that the chemical reaction is the rate determining step and the partial pressure of carbon dioxide is low.

Dr. Zavitsanos: In the case of coal combustion (surface oxidation step) did you observe carbon monoxide as one of the products; if so, what was its magnitude in relation to carbon dioxide?

Dr. van Heek: Until now we have measured only the carbon dioxide formation during the combustion, but it is known that carbon monoxide is primarily formed on the surface and oxidized to carbon dioxide in the adhering gas film. Measurements of the concentration profile of carbon monoxide near a burning graphite surface are published by E. Wick and G. Wurzbacher (*Int. J. Heat Mass Transfer* **5** 277/89 (1962).

Chapter 3

Multiple Pulse Radiolysis Studies of Inorganic Molecules with Low Energy Electrons

K. O. Dyson

Department of Physical Chemistry, University of Cambridge, England

Introduction

  The chemical effects of ionising radiation have been attributed mainly to (a) ionisation, (b) excitation, and (c) dissociation of the target molecules. In gases the chemical reactions observed appear to be independent of the type and energy of the ionising radiation. The processes responsible for the energy loss are, qualitatively, the same whether electrons, alpha particles or other ions are used as the ionising agents. The magnitude of the effects in a particular chemical system is determined by the total energy absorbed. At the molecular densities normally used in radiation chemistry, of the total ionisation produced, only about 25% can be accounted for by primary electron impact ionisation; the remainder is due to secondary or further reactions.

  Mass spectrometer ion sources have been extensively used in recent years to study these secondary reactions. This has been done by increasing both the pressure and the residence time of the ions in the source region. Primary ionisation has been studied since the earliest days of mass spectrometry; it would appear that the technique is providing information which is of real interest to the radiation chemist. This information is somewhat restricted to that obtained from the study of the initial and subsequent behaviour of ionic species formed in the ion source.

  There has been great interest in the excitation of molecules by electron impact. Mono-energetic electron beams have been used to study the energy range from the onset of excitation to beyond the ionisation potential. The resultant information is, cumulatively, leading to a greater understanding of this important energy region.

  Dissociation studies have been mainly concerned with the variation of the cracking pattern with ionising energy and other experimental variables. Classical mass spectrometric techniques have been most frequently used to examine samples after irradiation by external or more recently, internal sources of radiation. Dynamic mass spectrometry has enabled studies to be made of the ionic behaviour immediately after dissociation; studies of the self-decomposition of metastable ions are being reported with increasing frequency in the literature.

  This concentration upon the results of electron impact reactions with molecules has therefore been upon the ionic products. However there are also neutral products resulting from the ionisation radiation, particularly when the energies involved are greatly in excess of the molecular ionisation potential. It may be instructive to present an uncomplicated account of the reactions in an ion source so that a method may be developed of observing these neutral products without making major modifications to the physical layout of the mass spectrometer. The analysis which follows is not intended to be original nor is it an attack on the powerful theoretical aspects of mass spectrometry. The author's intention is to develop a simple model from which will ensue a logical path to a new instrumental technique.

Theory

Upon impact with a stable polyatomic neutral molecule an energetic electron can impart enough energy to the molecule to ionise it and also, by breaking one or more bonds, to produce fragment ions. Indeed many of the ions so formed may dissociate in turn to produce further fragment ions. The characteristic mass spectrum produced in this manner is the sum total of these separate ionising reactions; the total ionisation cross-section is equal to the sum of the specific ionisation cross-sections for each ion at the selected electron energy. Under some conditions the pressure-residence time product of the ions in the source region may be large enough for further ions to be formed, either by secondary reactions between primary ions or excited fragments with the parent molecule or by contributions from cumulative electron impact effects upon one or more species. Steps are usually taken to ensure that the number of ions formed from such secondary processes is negligible by comparison with primary ionisation.

The total number of ions formed each second in the source is M^+ where

$$M^+ = \sigma_{M^+} + M_o N_e l \tag{1}$$

where σ_{M^+} is the total ionisation cross-section for electrons of a selected energy, in cm² units, M_o is the density of the parent molecules, in molecules per cm³, N_e is the ionising current, in electrons per second and l is the effective path length of the electrons in the ion source, in cm.
Further

$$M^+ = P^+ + F^+ + S^+ \tag{2}$$

and
$$\sigma_{M^+} = \sigma_{P^+} + \sigma_{F^+} + \sigma_{S^+} \tag{3}$$

where P^+ is the number of primary parent ions, F^+ is the number of all of the primary fragment ions and S^+ is the number of all of the secondary ions, each formed per second. The ionisation cross-sections are expressed similarly in Eqn. (3).

Consider one specific fragmentation process,

$$M + e \rightarrow F_a^+ + R_a + 2e \tag{4}$$

where the parent molecule will produce a neutral species, R_a, in addition to the fragment ion F_a^+. The ionisation cross-section for Eqn. (4) will apply to F_a^+ from M ionisation and also to R_a from M production; thus a value may be assigned:

$$\sigma_{F_a^+} = \sigma_{R_a^+} = \frac{F_a^+ \; \sigma_{M^+}}{M^+} \qquad (5)$$

where F_a^+ is the abundance of the specific fragment ion and R_a the abundance of the resultant neutral.

A multiple pulse technique has been developed which permits an initial electron radiolysis pulse to be followed sequentially by ion clearance and ionising analysis pulses. This means that a short burst of radiation ionises the parent molecules and produces fragment ions: the existence of fragment ions also means that fragment neutral species are present in the volume element of the source region just vacated by the electron beam. The ions are removed and another burst of radiation permits further ionisation to take place. Provided that the time delay between the ionising pulses is short, the neutral fragment concentration produced during the radiolysis pulse will not decay to a great extent. The analysis pulse will then produce a normal spectrum of the parent molecule plus a contribution from ions formed from the neutral fragments. This extra contribution may be evident either as an increased abundance of some of the m/e ratios in the normal mass spectrum of the parent or as new m/e ratios in the cracking pattern; both effects may occur together.

That is,
$$M + e \rightarrow M^+ + 2e$$

and
$$M + e \rightarrow F_a^+ + R_a + 2e$$

produced by the radiolysis pulse; if all the charged species are removed by the ion clearance pulse, leaving $M + R_a$, then

$$M + e \rightarrow M^+ + 2e$$

$$M + e \rightarrow F_a^+ + R_a + 2e$$

$$R_a + e \rightarrow R_a^+ + 2e$$

is produced by the analysis pulse.

By analogy with Eqn. (1), the relationships governing the production and ionisation of a species R_a, in a way that is instrumentally useful, by means of such sequential reactions becomes

$$R_a^+ = \sigma_{F_a^+} \; \sigma_{R_a^+} \; [M] \; N_{er} N_{ea} l^2 \beta \qquad (6)$$

where N_{er}, N_{ea} are the densities, in electrons per second, for the radiolysis and analysis pulses respectively, β is a term describing the mass dependent, and possibly energetically dependent, decay in $[R_a]$ during the delay time between the electron pulses. An experiment will be described in which the preceding logic has been tested and has produced encouraging results.

Pulsed source mass spectrometers were not developed seriously until the last two decades when Wiley [1,2] applied the necessary techniques in order to produce a mass spectrum by mean of linear velocity analysis alone. Subsequent developments have concentrated on the application of pulse techniques in the ion source to the problems of ion focussing and withdrawal to the analyser. 'In source' velocity analysis and ion residence time studies, for probes into ion-molecule reactions, have been published recently by Henchman et al.[3-5]

The abrupt change in the chemistry of a sample that may be wrought by pulse techniques has been demonstrated in many fields. These range from the original flash photolysis works of Norrish, Porter and their associates[6] through the high energy pulse radiolysis studies reviewed by Dorfman and Matheson[7] to the low energy multiple pulse technique described in this paper. In all the preceding cases the pulsing technique was of the 'go-no go' variety; the energy source was effectively switched into either the ON or the OFF condition. It seemed a reasonable step to apply pulse techniques to the energy parameter in order to influence the course of ionising reactions.

Reed and Takhistor [8] have studied the variation in the cracking patterns of some hydrocarbons with manually adjusted electron energies. They have shown how the energy parameter controls the reaction scheme in successive fragmentation reactions. Their work was convincing evidence that the electron energy was a useful experimental variable for controlling the course of a reaction, in addition to the more traditional uses of the parameter. Melton,[9,10] in a series of papers ranging from 1959 to date, has described a powerful dual d.c. electron beam ion source which has been used to detect free radicals and metastables. In particular it was demonstrated that excited states could be produced in one beam and subsequently ionised in the second beam. A variation of one of his experiments will be reported using one electron beam only.

Instrumentation and technical details

The mass spectrometer used for these experiments was a Bendix T.O.F. Mass spectrometer, Model 14-105, equipped with a dual channel output system which utilises model 321 analogue scanners. The original pulse chassis has been retained for routine

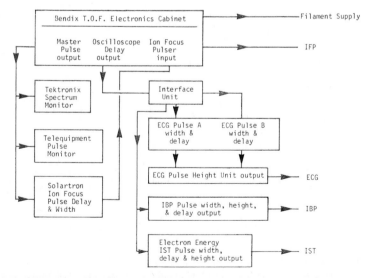

Fig. 1: Electronics block schematic for multiple pulse radiolysis ion source.

work, but for these studies a modified pulse system was assembled from commercial pulse generators and custom-built units.

The electronics block schematic is shown in Fig. 1 and from this it can be seen that every pulse could be specified for width and height. The time relationship of each pulse could be set from any other pulse or to the master pulse so that one or more outputs were delayed with respect to any preceding pulse. An Advance Instruments PG 52 unit was used to provide the pulses controlling the electron beam, electron energy and ion clearance parameters. Outputs of up to 20 V height, from 25 nsec width and below 5 nsec rise and fall times were available; the latter were checked on a Tektronix Sampling Oscilloscope for the appropiate termination. Typical settings were Control grid pulses of 0·1 to 1·0 µsec and ion clearance pulses of 0·2 to 1·0 µsec. The ion withdrawal pulse was controlled by a Solartron OPS.100C unit for width and delay; and by the Bendix Ion Focus output

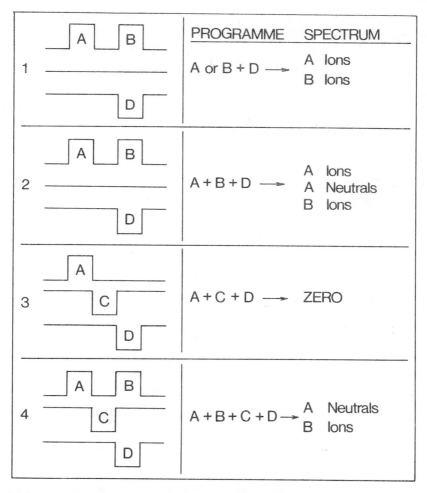

Fig 2: Multiple pulse radiolysis programme showing the four different pulse sequences and the respective spectrum contributions. A, B are electron beam control grid pulses; C is the ion clearance pulse; D is the ion focus (withdraw) pulse.

circuit for height. Custom-built units were used for various delay and ion repeller duties; generally the ion pulse specification was height 0-290 V, width 1 to 3 μsec and rise time 20 to 100 nsec.

The programme used for the multiple pulse radiolysis studies is shown in Fig. 2; this also shows the checks that were made for consistency of operation of the method. The pulse width C (backing plate pulse) is mass dependent and can be adjusted to remove the low mass ions only or the complete ion spectrum produced by the radiolysis pulse.

Various ion source designs were used ranging from the normal Bendix Wiley source to modified versions using restricted apertures and enclosed ionisation regions to enable higher source pressures to be achieved. A sketch of one source used is shown in Fig. 3 where the shaped backing plate and also the pulse programme used in the pulsed electron energy experiments can be seen. Other considerations limited the design criteria for this source and the resolution of the mass spectrometer using this source was much lower than is normally the case with the Wiley design. Figure 4 shows that the performance was perfectly adequate for these studies and a scan of the krypton isotopes resulted in a resolution figure of 70-80 using the 10% valley definition. The ion source tube (IST) was referred to ground and could be pulsed to control the electron energy and to collect or repel ions. The ion backing plate (IBP) and the ion focus grid (IFG) were also referred to ground and could be pulsed for ion transport.

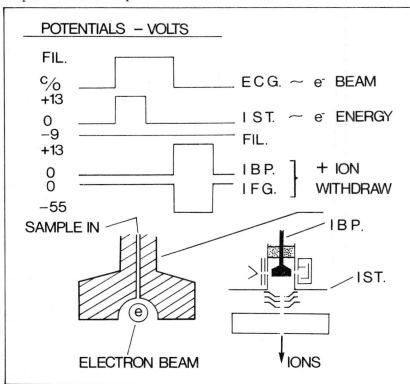

Fig. 3: Pulsed electron energy programme and sketch of modified source elements. ECG : Electron Control Grid; IST : Ion Source Tube; FIL : Filament; IBP : Ion Backing Plate, or Repeller; IFG : Ion Focus Grid, or withdraw electrode.

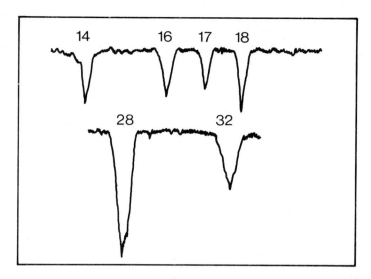

Fig 4: Resolution of the modified source of Fig. 3 over the m/e range used in this work.

Pressure measurements were made with thermal ionisation gauges or by total ion collection methods. A McLeod gauge was used as a calibrating reference and agreement to within 10% was considered satisfactory.

Standard vacuum preparative techniques were used for sample handling; anhydrous ammonia gas was obtained from an ICI Ltd. cylinder and distilled between dry ice and liquid nitrogen temperatures, middle fractions only were used. Helium was obtained from a British Oxygen Company cylinder and stored over liquid nitrogen to remove trace hydrocarbon impurities.

Results and discussion

Multiple Pulse Radiolysis. The ionisation cross-sections for transient neutral species are not known to the same precision as those of stable molecules; in many cases estimates have to be made by comparison with somewhat similar species. Melton[10] has described the relationship used to estimate the ionisation cross-section for neutral free radicals observed in a triple beam source as

$$\sigma_i = \sigma_o (E - I)$$

where σ_i is the ionisation cross-section for the selected value of the electron energy E, I is the ionisation potential of the molecule and σ_o is a constant.
For example, to obtain σ_i for NH_2 radicals, σ_i was determined for the parent ammonia molecule at a value for $(E-I)$ of 4 eV and was assumed to be equal to that for NH_2 at an equivalent $(E-I)$. As a working hypothesis Melton's technique has proved invaluable. It is of interest to obtain free-radical ionisation cross-sections by more direct means.

As an example, the multiple pulse radiolysis method was used to obtain the ionisation cross-section for hydrogen atom radicals produced by 100 eV radiolysis of ammonia. Previous studies by electron radiolysis[10] and electrical discharge techniques[11] had revealed that the two most abundant radical species produced during the decomposition of ammonia are NH$_2$ radicals and hydrogen atoms. It was assumed that the only process giving rise to the radical signal was

$$NH_3 + e \rightarrow NH_2^+ + H + 2e \qquad (7)$$

Contributions from

$$NH_3 + e \rightarrow NH_2^- + H$$

and
$$NH_3 + e \rightarrow N^+ + H_2 + H + 2e$$

were considered to be negligible by comparison with Eqn. (7). An ion source concentration was selected which yielded a good radical signal, but was sufficiently low to avoid interference from ion-molecule reaction products.

The observed difference signal current, divided by the experimentally determined multiplier gain, was inserted into the re-arranged Eqn. (6), thus;

$$\sigma_{iH^+} = \frac{\Delta H^+}{\sigma_{F^+} [M] N_{er} N_{ea} l^2 \beta}$$

where the conditions were:
σ_{iH^+} is the cross-section for ionisation of hydrogen atom radicals,
ΔH^+ is the current = 2×10^{-11} A = 1.25×10^3 ions sec^{-1}
[M] = 6×10^{11} molecules cm^{-3}
N_{er} = 6.3×10^{11} electrons sec^{-1}
N_{ea} = 6.3×10^{10} electrons sec^{-1}
σ_{F^+} = $\sigma_{NH_2^+}$ fragment ion from NH_3 = 0.8×10^{-16} cm^2, [10] l = 2 cm, β = 3 for a 1 μsec delay between the electron beam pulses and a beam diameter of 1 mm.

The value obtained, 5.4×10^{-17} cm^2, is compared in Table 1, with those presented by other workers using different techniques.

Table 1. Cross-section for the production of H from hydrogen atoms by 100 eV electrons. [a] ($\sigma \times 10^{16}$ cm^2 units)

Fite et al.[12]	Rothe[13]	Boksenberg[14]	This work
0.64	0.49	0.59	0.54

[a] Refs. 12, 13, 14 are conversions from $\bar{\sigma}(\pi a_0^2)$ units. References originate from a compilation supplied by Keiffer.[15,16]

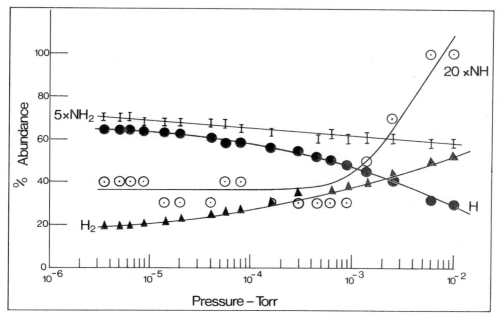

Fig. 5: Plot of neutral fragments v. ion source pressure for 100 eV radiolysis of ammonia.

Figure 5 shows that the hydrogen atom radical signal was abundant and provided a large signal. The figure also indicates that other neutral fragments were detected in significant concentrations. The changing abundances with increasing pressure indicates the probable participation of these neutrals in secondary reactions.[17] Three different ion sources with greatly differing characteristics were used to cover the total pressure range. The abundances are quoted as per cent per total neutrals detected. Absolute abundance variations with pressure await the construction of a suitable single source and sampling system which will enable a wide pressure range to be covered with constant efficiency and sensitivity. The reliability of measurement of the radical concentrations obviously depends on the difference between two signals; these are obtained with and without the radiolysis pulse. The hydrogen atom radical signal was the result of subtracting a small from a much larger signal. In these particular experiments the use of the NH_2 radical signal would have introduced a much greater error as that difference signal was a result of subtraction of one signal from another of the same order of magnitude.

Further development of the technique should enable the cross-sections for ionisation of other neutral species to be determined directly.

Pulsed Electron Energy experiments. A single electron beam may have its energy parameter pulsed from a pre-set bias level to a different, or a number of different, energy values in succession. These pulses are in addition to the pulse(s) that are used to switch the beam on or off. It is possible to devise experiments in which a molecule is required to be impacted by more than one electron and these may be of the same or different energies. One such experiment involved the raising of helium atoms to an excited, metastable level by a primary pulse followed by ionisation from low energy electrons which were permitted to pass through the ion source for the duration of the control grid pulse. Electrons of

21-24 eV (I.P. of helium 24 6 eV) should raise the atom to the $1s2s$ metastable state; low energy electrons should then be able to ionise the metastable atom.

The experimental proceedure adopted was to admit helium to the ion source and obtain stable conditions throughout the system. An I.P. run for He$^+$ was obtained and the ionisation onset region examined and compared to previous calibration plots for argon and helium. This has been found to be a necessary proceedure when one wishes to work at electron energies close to ionisation potentials as source characteristics vary with sample history, electron densities and simply with running time. Contact potentials are not fixed values to be determined at one time and to be applied as a calibration on all future occasions. It is most important to operate at the electron densities and filament temperature for which the calibration plots were obtained as the energy spread variation could cause irreproducible effects. The electron energy was then set to 25 eV and the He$^+$ signal monitored.

The control grid pulse, originally set for zero delay with respect to the ion withdrawal pulses, was slowly moved earlier in time and the delay increased to a point where the He$^+$ signal decreased to noise level. The electron energy was then reduced to 9 V at which a previously determined trap current was obtained and the noise level became minimal. A 13 V pulse was then applied to the ion source tube for the duration of the control grid pulse, so that the effective electron energy became 22 V for this period; the trap current increased by about 10% and no He$^+$ signal was observed. The control grid

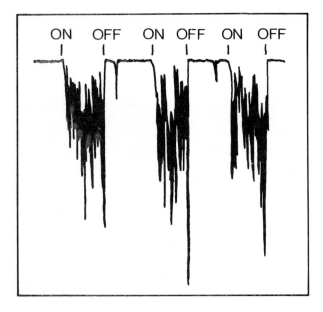

Fig. 6: Tracing of metastable helium signal. OFF indicates no IST pulse and the electron energy is therefore set by the filament level below ground (9 V). ON indicates that a 13 volt positive pulse to IST raises the electron energy to 22 volts. On cessation of the IST pulse the electron energy reverts to 9 volts for the remainder of the electron beam duration, which is controlled by the ECG pulse. Since both energy states of the electron beam were lower than the demonstrated laboratory ionisation of He, He$^+$ ions had to be formed by multiple electron impact.

pulse duration was then adjusted so that it covered the period from the start of the electron energy pulse to the start of the ion withdrawal pulses. The He$^+$ signal increased from the noise level; when the electron energy pulse was removed the energy reverted to 9 V and the He$^+$ signal, dropped to zero.

Figure 6 shows a recorder tracing obtained for the OFF and ON conditions of the electron energy pulse, Fig. 7 shows the laboratory conditions of the experimental para-

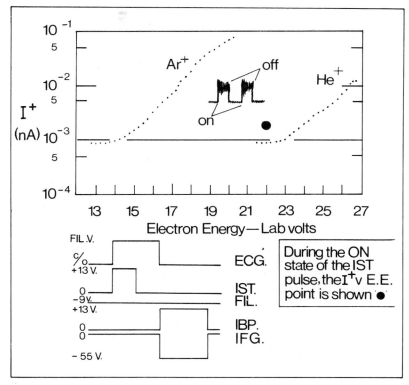

Fig. 7: Analogue output current for Ar$^+$ and He$^+$ versus laboratory electron energy; experimental conditions for the detection of the metastable helium signal.

meters. Any explanation for the appearance of the He$^+$ signal must preclude direct ionisation by single electron impact because the laboratory electron energy had been demonstrated to be insufficient for this purpose. It is reasonable to assume that 22 V electrons were producing helium metastable states which were subsequently ionised and trapped in the space charge of the low energy electron beam. It was not possible from this experiment to decide whether 22 V or 9 V electrons were responsible for the subsequent ionisation.

Several workers[18-24] have shown that ions may be stored for periods of up to 10^{-3} seconds in this manner. Many experiments now become possible in which excited species, neutral fragments and positive ions may be formed and subjected to further electron impact at closely controlled, pulsed electron energies whilst trapped in the valley potential of a low energy electron beam. Where the molecular weight of the fragment or excited neutrals is such that the residence time in the volume element mapped out by the electron beam is greater than the duration of the initial energy step or radiolysis pulse, then the subsequent energy-selected ionising beam is capable of supplying the trapping valley potential with the required ions for a considerable period after the cessation of the initial radiation. In this way the numbers of ions which may be obtained from the selected neutrals increases to a limiting value set by the velocity of the neutral species under the experimental conditions. The efficiency of detection, then, is higher than for multiple beam methods. When the 'solid angle of acceptance' factor for transfer of neutrals from one beam downstream to another is considered as a weighting factor then the arguments for using one beam for radiolysis and analysis become even more convincing.

Experiments are in progress [25] where it is possible to produce fragmentation at threshold energies which are well below positive ionisation potentials. Subsequent ionisation of these fragments occurs at electron energies which are intermediate to their ionisation potential and the appearance potentials of the similar m/e fragment ions from the parent molecule.

Acknowledgements

This work has been carried out during the course of development of the Cambridge T.O.F. mass spectrometer and is subsidiary to the main research programmes. I wish to record my appreciation to Professor J. W. Linnett for this facility and to Dr. P. F. Knewstubb for many kindly criticisms.

References

1. W. C. Wiley, *U.S. Pat.* 2,685,035 (1950).
2. W. C. Wiley and I. H. McLaren, *Rev. Sci. Instr.* **26**, 1150 (1955).
3. K. Birkinshaw, A. J. Masson, D. Hyatt, L. Matus, I. Opauszky and M. J. Henchman, *Advances in Mass Spectrometry,* Vol. 4, The Institute of Petroleum, London, p. 379.
4. K. Birkinshaw, A. J. Masson, D. Hyatt, L. Matus, I. Opauszky and M. J. Henchman, *Discussions Faraday Soc.* **44**, 146 (1967).
5. L. Matus, D. J. Hyatt, M. J. Henchman, *J. Chem. Phys.* **46**, 2439 (1967).
6. P. G. Ashmore, F. S. Dainton and T. M. Sugden (Eds.), *Photochemistry and Reaction Kinetics,* The University Press, Cambridge, 1967.
7. L. M. Dorfman and M. S. Matheson, *Progress in Reaction Kinetics,* Vol. 3, 1965, p. 239. (Ed. G. Porter) Pergamon Press, Oxford.
8. R. I. Reed and V. V. Takhistor, *Tetrahedron* **23**, 4425 (1967).
9. C. E. Melton, *J. Sci. Instr.* **43**, 927 (1966).
10. C. E. Melton, *J. Chem. Phys.* **45**, 4414 (1966).
11. K. O. Dyson in *Time-of-Flight Mass Spectrometry* (Eds. D. Price and J. Williams), Pergamon Press, Oxford, 1969.
12. W. L. Fite and R. T. Brackman, *Phys. Rev.* **112**, 1141 (1958).
13. W. Rothe, *Phys. Rev.* **125**, 582 (1964).
14. A. Boksenberg, *Ph.D. Thesis,* University of London, 1961.
15. L. J. Keiffer and G. H. Dunn, *Rev. Mod. Phys.* **38**, 1 (1966).
16. L. J. Keiffer, private communication.
17. G. Derwish, A. Galli, A. Giardini-Guidoni and G. Volpi, *J. Chem. Phys.* **39**, 1599 (1963).

18 A. J. Bourne, C. J. Danby, *J. Phys. (E).* **1**, 155 (1968).
19 N. R. Daly and R. E. Powell, *Phys. Rev. Letters* **19**, 1165 (1967).
20 N. R. Daly and R. E. Powell, *Proc. Phys. Soc. (London)* **89**, 273 (1966).
21 P. A. Redhead, *Can. J. Phys.* **45**, 1791 (1967).
22 J. Cuthbert, J. Farren and B. Prahallada Rao, *Proc. Phys. Soc. (London)* **88**, 91 (1966).
23 J. Cuthbert, J. Farren and B. Prahallada Rao, *Proc. Phys. Soc. (London)* **91**, 63 (1967).
24 F. A. Baker, J. B. Hasted, *Phil. Trans. Roy. Soc.* **A261**, 33 (1966).
25 K. O. Dyson, to be published.

Discussion

Dr. G. W. F. Pike: Can you explain how you arrived at a theoretical value for your time decay factor β? It would seem in part to depend on the cross-sections you wish to measure.

K. O. Dyson: As you obviously appreciate, β is a time and mass-dependent factor. At the end of the radiolysis pulse there will be a certain concentration of neutrals. For the duration of the ion clearance pulse this concentration will be decreasing with time as the thermal velocity of the neutral fragments permits them to move out of the volume element, V_o, originally occupied by the electron beam. Calculations based upon the r.m.s. velocity of the neutral fragments at the measured source temperature and the dimensions of the electron beam enable one to construct a simple model of an expanding volume element, V_t. The ratio of V_t to V_o is related to the neutral fragment density decrease and provides the time decay factor β.

There remains the question of the validity of inserting a constant based on this purely theoretical consideration into Eqn. (6). Harrison *et al.*,[1] in preliminary work on ion-molecule reaction studies, showed that the dependence of ion concentration with time, in their field-free ion source, followed a theoretically predicted model based upon the same tenets that I have used above. Ion concentration studies versus residence time in a 'field-free' Bendix type source (i.e. one in which steps were taken to remove field penetration effects) also show that similarly shaped curves are obtained for both theory and experiment. Therefore, I feel it is valid to use calculations that are, effectively, based upon the kinetic theory of gases.

Finally, I cannot see how the ionisation cross-section for a neutral fragment, or indeed any molecule, will affect its thermal velocity. It is this latter parameter that is important.

Reference
1 Harrison *et al., Can. J. Chem.* **41**, 3119 (1963); **43**, 159 (1965); **44**, 1351 (1966).

Chapter 4

Flash Photodecomposition of Lead Tetramethyl Monitored by Time-Resolved Mass Spectrometry

S. E. Appleby, S. B. Howarth, A. T. Jones, J. H. Lippiatt,
W. J. Orville-Thomas, D. Price and (in part) P. Heald

Department of Chemistry and Applied Chemistry, The University of Salford, England

Introduction

Although lead tetramethyl is of great interest because of its anti-knock properties, surprisingly little is known of its decomposition reactions. Walsh and Ting-Man Li[1] have investigated the decomposition of lead tetramethyl in a static system in the pressure range 2-25 torr and at temperatures of the order of 300°C. Gas chromatographic analysis of the gaseous products showed the products, in descending abundance, to be ethane > methane > propane > ethylene. No solid carbon was observed and the carbon/hydrogen ratio was 1/3. The relative product ratios remained constant up to 50% decomposition. Addition of toluene was found to increase methane production at the expense of ethane. This was taken as being indicative of the presence of methyl radicals in the reaction. The reaction was shown to be homogeneous. On the basis of their results, Walsh and Ting-Man Li proposed that the following reactions are involved in this decomposition reaction:

$$Pb(CH_3)_4 \rightarrow CH_3 + Pb(CH_3)_3$$

$$CH_3 + Pb(CH_3)_4 \rightarrow CH_4 + CH_2Pb(CH_3)_3$$

$$CH_3 + CH_3 \rightarrow C_2H_6$$

$$2 CH_2Pb(CH_3)_3 \rightarrow [(CH_3)_3PbCH_2CH_2Pb(CH_3)_3] \rightarrow C_2H_4 + 2 Pb(CH_3)_3$$

A single pulse shock tube was used by Ryason[2] to study the thermal stabilities of lead alkyls between 731°K and 931°K. He found the decomposition to be first order and the bond energy of the Pb — C bond in lead tetramethyl to be 34·9 kcal mole^{-1}.

Leighton and Mortensen[3] studied the photolysis of lead tetramethyl in the region of continuous absorption between 2000 Å to 2800 Å. The main product was ethane and the principal overall reaction proposed was

$$Pb(CH_3)_4 \rightarrow Pb + 2 C_2H_6$$

The quantum yield at 2537 Å was 1·11. The presence of free radicals was indicated in experiments with radioactive lead tetramethyl. Clouston and Cook[4] flash photolysed lead tetramethyl with 2,500 joules energy to obtain greater than 90% decomposition under adiabatic conditions in the pressure range 1-10 torr. Absorption spectroscopy was used to monitor the system, but quantitative data could not be obtained because the optical windows became opaque within 1 millisecond of the flash. The absorption spectrum was interpreted as indicating the presence of a free radical of formula $Pb(CH_3)_x$ where $x = 1$, 2 or 3. Infrared analysis showed the products to be ethane, ethylene, methane and acetylene. The mechanism proposed was

$$Pb(CH_3)_4 \xrightarrow{h\nu} CH_3 + Pb(CH_3)_3$$

$$Pb(CH_3)_3 \xrightarrow[\text{rapidly}]{\text{very}} Pb + 3\,CH_3$$

so that for all practical purposes the overall reaction is

$$Pb(CH_3)_4 \xrightarrow{h\nu} Pb + 4\,CH_3$$

The work of Clouston and Cook is typical of previous attempts to investigate the decomposition of lead tetramethyl using optical techniques. These have been handicapped by the rapid deposition of metallic lead onto the windows, thus preventing optical measurements from being made. The big advantage of using time-resolved mass spectrometry is that such fogging does not affect the quantitative accuracy of the mass spectrometric measurements. Thus, a successful investigation of this reaction would well illustrate the value of developing the time-of-flight mass spectrometer to monitor flash-photolysed reactions.

The technique of flash photolysis as pioneered by Norrish and Porter is well known.[5] The intense photoflash produces high transient concentrations of reactive species and can thus initiate fast reactions. Hitherto the most common method of monitoring flash-photolysed reactions has been absorption spectroscopy, using either a spectrograph to obtain a complete absorption spectrum or a monochromator to follow the change in absorption at one particular wavelength characteristic of a species of interest.

Absorption spectroscopy can give both qualitative and quantitative data concerning a flash-photolysed reaction within the very short time scales involved. The method also has the advantage that it does not disturb the system. However, many important reactive intermediates cannot be detected by their adsorption spectra because these are either too weak or are continuous. Thus the development of another method of monitoring flash-photolysed reactions would enlarge the range of chemical reactions which can be studied. Because of its ability to give both qualitative and quantitative data, mass spectrometry[6] is an obvious possibility as a monitoring technique although it is probable that only gas phase reactions can be monitored.

The magnetic deflection instrument has been little used in this respect although Strausz et al. have used it to study the flash photolysis of carbonyl selenide[7] and tricarbonylcyclobutadienyliron.[8] The disadvantage of their system is that only one peak in the mass spectrum of the reaction system can be followed in any one experiment. Of the dynamic mass spectrometers, the quadrupole has been used by Capey[9] to investigate the flash photodecomposition of nitrogen dioxide but a greater effort has been put into the development of the time-of-flight mass spectrometer.[10] The latter instrument, with its ability to produce a complete mass spectrum every 10^{-4} seconds or less, would seem to be the best type of mass spectrometer for this purpose. In fact, the stimulus for the development of the time-of-flight mass spectrometer came from the efforts of Kistiakowsky and Kydd[11] to use it to study the flash photolysis of nitrogen dioxide and ketene.

The major problem in the application of the time-of-flight mass spectrometer to monitor fast reactions is due to its pulsed mode of operation. Because any mass peak on a given trace results from a finite and often small number of detected ions, statistical fluctuations occur. These fluctuations are superimposed on any change of concentration due to chemical reaction. Fluctuations of this general type tend to be noticeable in any experiment in which data have to be obtained in a very short time interval. To minimise the effect of these fluctuations it is essential to produce the maximum possible number of ions per cycle of the time-of-flight mass spectrometer. Meyer[10] has successfully shown that such an instrument can be successfully utilised to monitor flash-photolysed gas reactions. In particular, he has detected the presence of the hydroxyl radical in the nitrogen dioxide photosensitised reaction between hydrogen and oxygen[12] and investigated the flash photolysis of methyl iodide.[13,14]

This chapter describes the development of a flash photolysis/time-of-flight mass spectrometry system and its application to study the flash photodecomposition of lead tetramethyl.

Flash photolysis time-of-flight mass spectrometer apparatus

The overall experimental arrangement is shown in Fig. 1. It is best described section by section.

The Mass Spectrometer is a Bendix Model 14-107 fitted with a fast reaction chamber. Several modifications were made to the original instrument in order to maximise the number of ions produced, transmitted and collected per cycle. A brief description follows.

(i) Tightly stretched molybdenum mesh grids were used in the ion source.
(ii) The electron beam pulse duration was increased from 0·25 to 1·4 microseconds.
(iii) Time-lag focussing[15] was incorporated both to improve resolution and because it is essential to prevent overlap of the electron beam and ion focus pulses when the former is increased to 1·4 microseconds duration. This modification affects the level of the trigger signal from the mass spectrometer to the oscilloscope. It is essential to allow for this effect.
(iv) An intermediate accelerating potential of −700 volts was applied to the second ion grid.[16]
(v) The ion source was differentially pumped by means of a separate 6″ diffusion pump system in order to minimise the effect of background gases and collisions involving sampled species.
(vi) Other instrumental parameters were adjusted to maximise the ion intensities for each set of experimental conditions and for the mass range of interest.

The instrument was operated at 20 kHz with the electron beam energy at 70 eV and an indicated trap current of 25 μA. The number of ions per cycle of the modified instrument was determined by measurement of argon ion statistics. The calculation was made in the same manner as those of Kistiakowsky[11] and Meyer[17] and showed that approximately 50,000 ions per cycle were produced in the modified instrument. Since single ions are

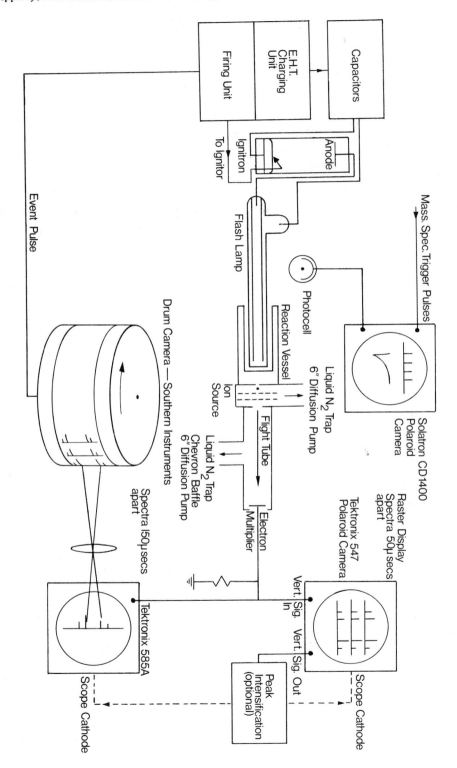

Fig. 1: Schematic of flash photolysis-time-of-flight mass spectrometer apparatus.

detectable, this number of ions provides a relative sensitivity of 1 part in 50,000 for a species having a single peak mass spectrum. For a species with more than one peak in its mass spectrum then the relative sensitivity will be given by 500 x % abundance of the particular ion chosen to detect the species. In concentration terms, for a reaction pressure of 10 torr, the lowest detection limit will be 10^{-6} to 10^{-7} mole l^{-1} for a species with a single peak mass spectrum and somewhat higher for a species which yields ions of more than one m/e ratio in the mass spectrometer.

Data Recording. The output from the electron multiplier was fed to the two oscilloscopes as shown in Fig. 1. Both oscilloscopes had P11 phosphors. The initial six or so spectra after the flash were displayed on the 547 scope using a raster circuit.[18,19] These spectra are photographed using a Polaroid camera fitted with an f1·4 lens and loaded with Polaroid 410 (10,000 A.S.A.) film. The second scope (585A) was set to display the output of every third cycle of the mass spectrometer, i.e. at 150 μsec intervals. This gave a spacing of 3/8" between spectra photographed with the drum camera at its highest speed. This camera photographed about 50 spectra before the flash and a simlar number immediately following the flash. The film used was 70 mm Kodak 2485 developed in Kodak MX 642-1 for 10 minutes at 20°C (nominal rating 10,000 A.S.A.). Attempts to cut the development time by using higher temperatures tended to diminish the contrast obtained and were therefore, abandoned.

This combination of oscilloscopes and cameras was able to record a mass spectrum every 50 μsec during the period of rapid initial reaction and, thereafter every 150 μsec for about 10 msec when the reactions were slower. The good correlation between the two recording techniques is illustrated in Fig. 2.

The main problem in photographing mass spectra displayed on an oscilloscope screen is that the peaks are very narrow pulses ($\sim$ 20 nsec) on a relatively slow time base sweep ($\sim$ 0·5 μsec cm^{-1}). If the scope intensity is turned up sufficiently for the taller pulses to be visible, then the base line becomes over accentuated and hence small peaks may be lost. Also, the diameter of the spot formed by the oscilloscope electron beam becomes comparable in size to the heights of the small peaks, rendering measurement of the latter non-quantitative. This affects the available dynamic range. The 'peak intensification' technique[20-23] is designed to offset this effect by intensifying the scope beam as it traces a peak compared to when it traverses the base line. This enables the base line intensity to be reduced greatly relative to that of the peaks. This can be effected either by amplifying the vertical signal out of the 547 scope and then applying the amplified signal as a negative voltage to the scope cathodes or by using the vertical signal to trigger a pulse generator which then provides the negative signal to the scope cathodes. The advantage of the latter method is that it provides constant intensification. The disadvantage is that small signals may be insufficient to trigger the pulse generator and are, therefore, not intensified. We are in the process of experimenting with both techniques. The data described in this paper were obtained without use of peak intensification.

The drum camera also functioned as the initiator of the whole experiment by acting as the on/off switch in the circuit carrying the trigger signal from the mass spectrometer to the 585A oscilloscope. Initially, this trigger signal was switched off. When the complete experiment had been set up and the drum camera was at maximum speed, a trip switch on the camera was operated. The trigger signal to the scope was then switched on for a single revolution of the drum thus preventing overlapping of the spectra. Halfway through this revolution, an event pulse from the drum camera initiated firing of the flash lamp via the

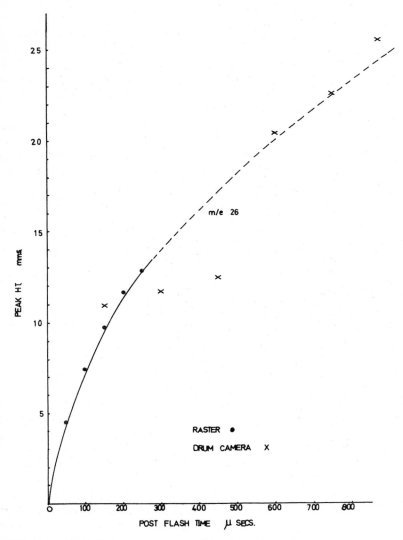

Fig. 2: Joint plot of raster and drum camera film data.

firing unit and the ignitron.[24] The raster circuit on the 547 scope was activated by the same pulse.

Timing. The cycling of the mass spectrometer and the initiation of the flash were not synchronised. Therefore, it was essential to record the timing of the flash relative to the mass spectrometer pulses. Such a record was obtained by photographing the trigger pulses from the mass spectrometer and the signal (vs. time) from the photocell when both were simultaneously displayed on the dual beam, single time base Solartron CD 1400 oscilloscope (Fig. 1). Zero time was taken as the peak of the photocell trace. The time before the first mass spectrum was obtained after the flash was easily interpolated and thus the time scale of the monitored reaction obtained.

Flash Lamp and Reaction Vessel. The geometry of the quartz lamp and reaction vessel is

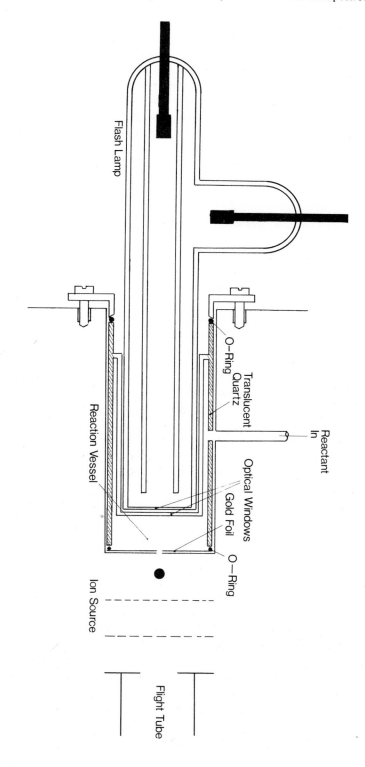

Fig. 3: Schematic showing the quartz flash lamp and reaction vessel geometry.

shown in Fig. 3. The reaction zone was essentially the volume, 1·5 cm x 2·5 cm diameter, between the optical window and the gold foil separating the reaction zone from the ion source. The reacting system was constantly sampled via the 0·0005" pinhole at the centre of the gold foil. The distance between the pinhole and the electron beam was approximately 4 mm. The outer translucent sleeve of the reaction vessel prevented flash desorption of surface contaminants into the ion source.[25] Reaction pressures were measured by means of a dibutyl phthalate manometer.

The flash lamp was filled with either krypton or xenon at 5 torr and gave a continuum between $\sim$ 2200 Å and $\sim$ 4400 Å when energised by 1000 joules. The light output of the lamp was determined by nitrogen dioxide actinometry[17] to be of the order of 10^{18} quanta per cm² per flash.

Experimental

Reagents. Lead tetramethyl (purity 99·99% as determined by gas chromatography) was donated by Associated Octel Ltd. Argon, krypton, xenon and neon were British Oxygen Co. research grade and were used directly from the cylinder.

Method. Various mixtures of lead tetramethyl with argon as internal standard and neon, krypton or xenon as diluents were flash photolysed in the pressure range 1-2·5 torr. The flash was of 1000 joules (20 μF at 10 kV) energy with a width at 1/e peak height of 30 μsec. The reaction cell was cleaned before each experiment to prevent diminution of the photolysis flash due to deposited lead. The peak heights of the photographed mass spectra were measured using a travelling microscope.

Treatment of data

The raw data obtained immediately after the flash cannot be used directly to calculate the results because[26] (i) electromagnetic interference from the flash lamp discharge perturbs the mass spectrometer response and (ii) flash heating of the gases in the open-ended vessel manifests itself in a density decrease which will cause the ion current to decrease from the real value.

Such systematic variations are accounted for by reference to the argon $m/e = 40$ peak which is used as the internal standard. Corrected peak height

$$h_{corr} = \frac{\bar{h}_{40}}{h_{40}} \times h_{meas}$$

where $\bar{h}_{40}$ = average value of argon $m/e = 40$ peak before the flash
h_{40} = measured argon $m/e = 40$ peak in the spectrum of interest
h_{meas} = measured peak height of the m/e peak of interest.

Figure 4 shows some typical raw data plots, while Fig. 5 shows these corrected by this procedure. The fluctuating pattern of the curves obtained is due to the statistical fluctuation in the total number of ions from cycle to cycle of the mass spectrometer and to the fluctuating share of these ions between the species present in the ion source. Since these fluctuations are statistical, the best fit smooth curves were drawn through the points which had been corrected for systematic errors. The result was a series of ion intensity (peak height) versus time curves for various m/e values which must be converted to concentration (or partial pressure) time curves for the various species in the reaction vessel.

The composite mass spectrum of the reaction mixture, obtained at any instant in time, can be used to calculate the concentrations of the various species present using the

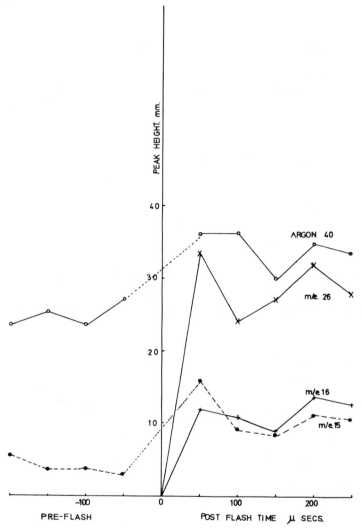

Fig. 4: Plot of uncorrected mass spectra data.

subtraction technique[27,28] provided the various cracking patterns and sensitivies are known. The species detected after decomposition of lead tetramethyl and the *m/e* peaks used for their analysis are given in Table 1.

Initially, the cracking patterns were obtained from separate experiments with a known pressure of a single species in the reaction vessel. Thus the absolute sensitivity (peak height per unit pressure) could be determined. However, these sensitivity values fluctuated so markedly that good analyses could not be obtained for known mixtures. Dove[29] has also observed this effect in sampling from a shock tube. The mass flow through the pinhole is dependent on the composition of the sample so the the response of the mass spectrometer will vary with the composition of the system being sampled. Therefore, all results described in this chapter were calculated using relative sensitivity values

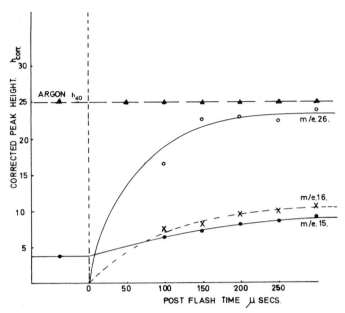

Fig. 5: Plot of data from Fig. 4 corrected for systematic variations.

Table 1.

Species	m/e peaks				
Hydrogen	2				
Methyl radicals	15				
Methane	15	16			
Acetylene	26				
Ethylene	26	27	28		
Ethane	26	27	28	29	30
C_3 hydrocarbons	42	43	44		

obtained from known mixtures of similar composition and pressure range to those present in the flash-photolysed reactions. The inert gas argon was used as reference. The basis of the method is best understood from consideration of a two-component system.

The mass spectrum peak height (h) is related to the partial pressure (P) of a species in the reaction zone by the expression $h = SP$ where S is the sensitivity. The relative sensitivity of the chosen peak of a species 'x' to that of the $m/e = 40$ peak of the reference argon, can be determined as

$$\frac{S_{Ar}}{S_x} = \frac{h_{Ar} \cdot P_x}{h_x \cdot P_{Ar}}$$

from the mass spectrum obtained with a mixture of known partial pressures, P_x and P_{Ar}, in the reaction vessel. In the subsequent kinetic studies the partial pressure of species 'x' can be calculated from the h_x/h_{Ar} ratio measured from the mass spectrum at any instant, the relative sensitivity and the known partial pressure of argon.

Species detected

The data were obtained from three different series of experiments in which the lead tetramethyl:diluent ratio was varied in order to observe the effect of changing the extent of adiabatic heating on the species produced by the flash photolysis. Figures 6, 8 and 9 show

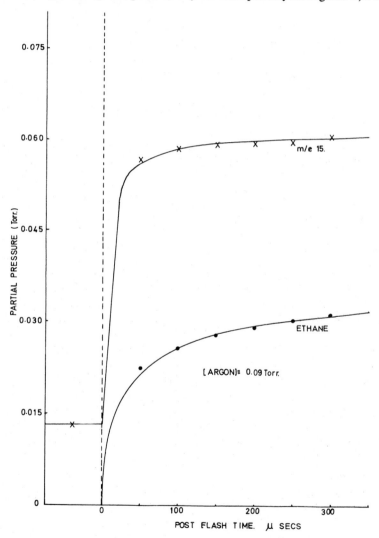

Fig. 6: Plot of partial pressure of products vs. time for Series A experiments (least adiabatic heating). Initial total pressure 2 torr; $Pb(CH_3)_4$: Ar : Ne = 1:0.6:12.4. The $m/e = 15$ curve represents CH_3^+ ions produced from lead tetramethyl and CH_3 radicals present in the reaction zone. Flash occurred at zero time.

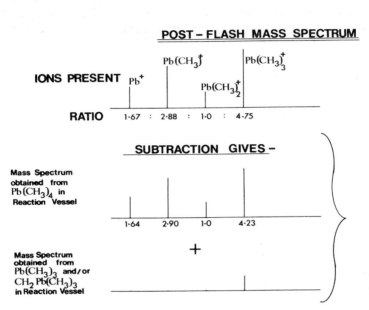

Fig. 7: Highest peaks due to Pb^+ and $Pb(CH_3)_x^+$ ions in mass spectra obtained during Series A experiments.

the various products detected during the first $400\mu sec$ after the lead tetramethyl had been flash photolysed.

Series A: Greatest dilution of lead tetramethyl – least adiabatic heating. Figure 6 shows the formation of the only two significant species – methyl radicals and ethane immediately after the flash. Little or no methane was detected and, therefore, the large m/e 15 peak increase must be due to presence of CH_3 radicals in the reaction zone. No unsaturated hydrocarbons were observed but traces of C_3 species were detected within $150\mu sec$ of the flash.

Figure 7 is a diagram of the highest peaks of the $Pb(CH_3)_x^+$ ions in the mass spectrum of this reaction system taken before and several seconds after the flash. The peak height ratios before the flash are characteristic of the lead tetramethyl mass spectrum cracking pattern. The singular 10% increase in the ratio value for the $Pb(CH_3)_3^+$ peak after the flash is indicative that this species was not then being produced solely by ionisation of lead tetramethyl. Other species from which the $Pb(CH_3)_3^+$ ion can be produced must be present in the reaction zone. The most probable species which may be postulated in this respect are $Pb(CH_3)_3$ and $CH_2Pb(CH_3)_3$ radicals. It is, therefore, proposed that the data are evidence for the presence of either or both of these species when lead tetramethyl is flash photolysed under the conditions of these experiments.

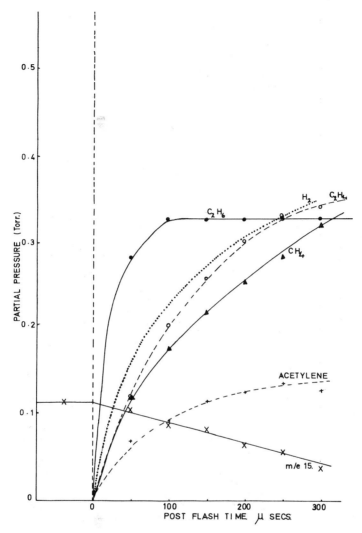

Fig. 8: Plot of partial pressure of products vs. time for Series B experiments (intermediate adiabatic heating). Initial total pressure 2 torr; $Pb(CH_3)_4$: Ar : Xe = 1:1:1. The $m/e = 15$ curve represents CH_3^+ ions produced from lead tetramethyl and any CH_3 radicals present in the reaction zone, contributions from other sources having been subtracted. Flash occurred at zero time.

This proposal, that the $Pb(CH_3)_3$ radicals are still present several seconds after the flash, contradicts the Clouston and Cook postulate[4] that $Pb(CH_3)_3$ is very unstable and essentially behaves as $Pb + 3 CH_3$. Their paper favours $Pb(CH_3)_2$ as the polymethyl lead species produced on flash photolysis of lead tetramethyl. The mass spectrum of lead tetramethyl shows that the $Pb(CH_3)_2^+$ ion is more unstable than the $Pb(CH_3)_3^+$ ion. It should be noted that Clouston and Cook used flash energies 2·5 times greater than those used in these experiments and it may be that in these circumstances $Pb(CH_3)_3$ is very unstable. At the same time, one would anticipate that $Pb(CH_3)_2$ would be even more unstable.

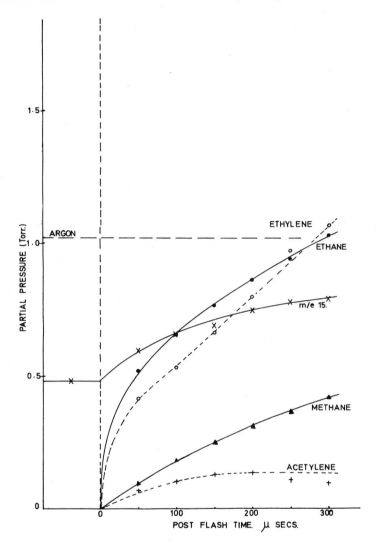

Fig. 9: Plot of partial pressure of products vs. time for Series C experiments (greatest adiabatic heating). Total initial pressure 2 torr; $Pb(CH_3)_4$: Ar = 1:1·2. The m/e = 15 curve represents CH_3^+ ions produced from lead tetramethyl and any CH_3 radicals present in the reaction zone, contributions from other sources having been subtracted. Flash occurred at zero time.

Series B: Intermediate dilution of lead tetramethyl – intermediate adiabatic heating. The curves shown in Fig. 8 are a composite representation of five separate experiments. No C_3 species were detected. No ions were detected in an experiment in which the electron beam was switched off; photoionisation therefore did not occur under the conditions of our experiments.

Series C: Least dilution of lead tetramethyl-greatest adiabatic heating. Fig. 9 shows the rates of formation of the various products in the 400 μsec immediately after the flash. In addition, a trace of hydrogen atoms was observed soon after the flash and much hydrogen was detected in the mass spectrum of the products some 10 sec or so after the flash. The failure to detect hydrogen molecules immediately after the flash, although ethylene and acetylene were present, was probably due to poor response of the instrument to $m/e = 2$ during these experiments. No C_3 species were detected.

The more interesting observations concerning Fig. 6-9 are summarised in Table 2.

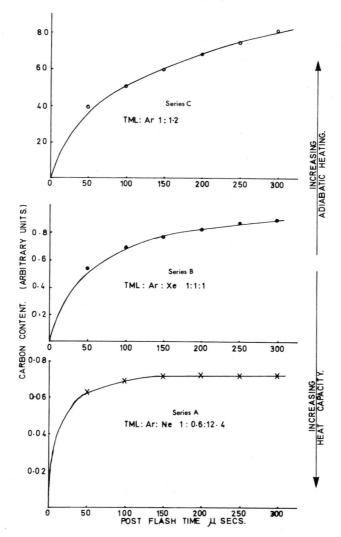

Fig. 10: Plot of carbon content of products produced by flash photolysis of lead tetramethyl vs. time showing the effect of variation in the extent of adiabatic heating.

Figure 10 shows the increase with time of the carbon content of the products for the three different experimental conditions. With the least adiabatic heating the carbon content increased rapidly during the first 50 μsec after the flash and, thereafter, remained constant. This suggests that in this case the lead tetramethyl is essentially photodecomposed during the flash and then takes very little further part in the reactions.

In the medium adiabatic heating case, the carbon content increases rapidly in the first 150 μsec and is still increasing 400 μsec after the flash. With the greatest adiabatic heating, the carbon content is still increasing rapidly after 400 μsec. These observations indicate that as the temperature of the reacting system is increased, there is an increase in the participation of lead tetramethyl in secondary reactions following the initial photo-decomposition.

Discussion of reaction mechanism

Primary Photodecomposition. All available experimental data indicate that the major primary processes must be

$$Pb(CH_4)_4 + h\nu \longrightarrow \begin{cases} CH_3 + Pb(CH_3)_3 & (a) \\ Pb(CH_3)_x + y\,CH_3 \text{ where } x+y=4 & (b) \\ Pb + 4\,CH_3 & (c) \end{cases}$$

Previous workers have postulated (a) and (c) as the major reactions. The detection of methyl and probably lead trimethyl radicals reported herein does not contradict these postulates. Some of the methyl radicals will be in excited states, the percentage of these increasing with increasing initial temperature rise, i.e. increase of adiabatic heating.

The amount of ethane present after 50 μsec in Series A (least adiabatic heating) was compared with the quantity calculated to be formed in the same time by the reaction $CH_3 + CH_3 \rightarrow C_2H_6$ using an average value for the reaction rate constant of 2×10^{13} cm^3 mole^{-1} sec^{-1} and assuming 20% initial decomposition of the lead tetramethyl [reaction (c)]. The calculation showed that the measured concentration of ethane could be completely accounted for by this reaction.
The reaction

$$Pb(CH_3)_4 + h\nu \rightarrow Pb + \tfrac{x}{2}C_2H_6 + y\,CH_3$$

where $x + y = 4$

does not appear to be important under these conditions. This does not rule out its occurrence under conditions of greater adiabatic heating but this would be more difficult to check since ethane is both being formed and removed under these conditions.

Secondary Reactions. Both saturated and unsaturated hydrocarbons were detected in these experiments. It is convenient to discuss their formation separately.

(a) *Formation of saturated hydrocarbons.* Ethane and methane were detected as major products in Series B and C. The principal reactions by which they were formed are probably

$$CH_3 + CH_3 \rightarrow C_2H_6$$

$$CH_3^* + CH_3 \rightarrow C_2H_6^* \xrightarrow{+M} C_2H_6$$

$$CH_3 + Pb(CH_3)_4 \rightarrow CH_4 + CH_2Pb(CH_3)_3$$

Gowenlock[30] has determined the value of the rate constant of the latter reaction.

This is given by the equation

$$k = 10 \cdot 2 \pm 0 \cdot 48 \, e^{\frac{-7,400 \pm 88}{RT}}$$

Another possible reaction is

$$CH_3^* + Pb(CH_3)_4 \rightarrow C_2H_6 + Pb(CH_3)_3$$

A study of the variation of the rate of ethane formation with the initial concentration of lead tetramethyl should indicate if this reaction is significant.

(b) *Formation of unsaturated hydrocarbons.* Ethylene and acetylene were detected as major products when the extent of adiabatic heating was markedly increased. Knox, Norrish and Porter[31] observed a similar increase in the extent of formation of unsaturated hydrocarbons with increase of adiabatic heating during a study of the photodecomposition of ketene.

Table 2 shows that there was a marked decrease in the ethane:ethylene ratio with increasing initial temperature rise. The effect of increasing temperature is, therefore, to increase the rate of production of ethylene as opposed to the rate of formation of ethane. This could be explained by an increased concentration of hot 'methyl' radicals. The most likely reactions are

$$CH_3^* + CH_3^* \text{ (or } CH_3) \rightarrow C_2H_6^*$$

$$C_2H_6^* \rightarrow C_2H_4 + H_2$$
$$ \rightarrow C_2H_4^* + H_2$$

$$C_2H_4^* \rightarrow C_2H_2 + H_2$$

In Figs. 8 and 9 the acetylene concentration is virtually constant after 200 μsec, whilst 400 μsec after the flash there is more ethylene present than ethane and the ethane:ethylene ratio is still decreasing. After about 300 μsec, any 'hot' species would be removed either by reaction or by collisional deactivation. Thus, ethylene and acetylene can no longer be produced by reactions involving these species. Since ethylene is still being produced after 400 μsec, then reactions not involving 'hot' species must be responsible. The known presence of methyl radicals and ethane in the system suggests that reactions such as the following

$$CH_3 + C_2H_6 \rightarrow C_2H_5 + CH_4$$

$$C_2H_5 + CH_3 \rightarrow C_2H_4 + CH_4$$

$$C_2H_5 \rightarrow C_2H_4 + H$$

may be occurring.

Table 2. Summary of observations made for data obtained during these investigations.

	Least Adiabatic Heating	Intermediate Adiabatic Heating	Greatest Adiabatic Heating
Series	A (Figs. 6 & 7)	B (Fig. 8)	C (Fig. 9)
At 50 μsecs			
Radicals detected	CH_3 $Pb(CH_3)_3$ and/or $CH_2Pb(CH_3)_3$		CH_3
Species in order of decreasing concentration	C_2H_6 only	$C_2H_6 > H_2 > C_2H_4 > CH_4 > C_2H_2$	$C_2H_6 > C_2H_4 > CH_4 > C_2H_2$
Ethane:Ethylene ratio		2·30	1·30
Ethane:Methane ratio		2·30	4·80
Rates		$\dfrac{d[C_2H_6]}{dt} > \dfrac{d[H_2]}{dt}$ $\simeq \dfrac{d[C_2H_4]}{dt} > \dfrac{d[(CH_4)]}{dt}$ $> \dfrac{d[C_2H_2]}{dt}$	$\dfrac{d[C_2H_6]}{dt} > \dfrac{d[C_2H_4]}{dt}$ $> \dfrac{d[CH_4]}{dt} > \dfrac{d[C_2H_2]}{dt}$
At 400 μsecs			
Radicals detected	CH_3 $Pb(CH_3)_3$ and/or $CH_2Pb(CH_3)_3$		CH_3
Species detected in order of decreasing concentration	C_2H_6 only	$H_2 \simeq C_2H_4 > CH_4 > C_2H_6 > C_2H_2$	$C_2H_4 > C_2H_6 > CH_4 > C_2H_2$
Ethane:Ethylene ratio		0·90	0·90
Ethane:Methane ratio		0·95	2·4
Rates		$\dfrac{d[H_2]}{dt} \simeq \dfrac{d[C_2H_4]}{dt} >$ $\dfrac{d[CH_4]}{dt}$ $\dfrac{d[C_2H_6]}{dt} \simeq \dfrac{d[C_2H_2]}{dt} \simeq 0$	$\dfrac{d[C_2H_4]}{dt} > \dfrac{d[C_2H_6]}{dt} >$ $\dfrac{d[CH_4]}{dt}$ $\dfrac{d[C_2H_2]}{dt} \simeq 0$

The experimental curves show that methane is still being formed after 400 μsec but not at a faster rate than ethylene, as would be expected from these equations. Another possible source of ethyl radicals and hence of ethylene are the $CH_2Pb(CH_3)_3$ species formed by hydrogen abstraction from lead tetramethyl. Possible reactions are

$$Pb(CH_3)_4 + CH_2Pb(CH_3)_3 \rightarrow CH_3CH_2 + 2\,Pb(CH_3)_3 \qquad (d)$$

$$(CH_3)_3PbCH_2 + CH_2Pb(CH_3)_3 \rightarrow [\,(CH_3)_3PbCH_2CH_2Pb(CH_3)_3\,] \qquad (e)$$
$$\downarrow$$
$$C_2H_4 + Pb(CH_3)_3$$

Table 2 shows that the rate of formation of ethylene is greater than the rate of formation of methane after 400 μsec. This would appear to add weight to the suggestion that the reactions (d) and (e) occur under these experimental conditions. The fate of the $Pb(CH_3)_3$ species is presumably that they eventually decompose to lead and methyl radicals.

Acknowledgements

The authors wish to thank Associated Octel Ltd. for the sample of lead tetramethyl and Shell Research Ltd. for financial assistance. Two of us (S.B.H. and P.H.) are indebted to the Science Research Council for maintenance grants.

References

1. D. E. Hoare, Ting-Man Li and A. D. Walsh, *Proc. 12 Intern. Symp. Combustion, Poitiers, France, 1968,* Combustion Inst., 1969, p. 357.
2. P. R. Ryason, *Combustion Flame* **7**, 235 (1963).
3. P. A. Leighton and R. A. Mortensen, *J. Am. Chem. Soc.* **58**, 448 (1936).
4. J. G. Clouston and C. L. Cook, *Trans. Faraday Soc.* **54**, 1001 (1958).
5. *Photochemistry and Reaction Kinetics* (Eds. P. G. Ashmore, F. S. Dainton and T. M. Sugden), Cambridge University Press, 1967.
6. *Time-of-Flight Mass Spectrometry* (Eds. D. Price and J. E. Williams), Pergamon Press, Oxford, 1969, p. 25.
7. H. E. Gunning, P. Kebarle, W. B. O'Callaghan, O. P. Strausz and W. J. R. Tyerman, *J. Am. Chem. Soc.* **88**, 4277 (1966).
8. H. E. Gunning, M. Kato, P. Kebarle, S. Masamune, O. P. Strausz and W. J. R. Tyerman, *Chem. Commun.* 497 (1967).
9. W. D. Capey (University of Essex, England) in a lecture given to Mass Spectrometry Discussion Group, December 1968.
10. R. T. Meyer and D. Price, *Time-of-Flight Mass Spectrometry* (Eds. D. Price and J. E. Williams), Pergamon Press, Oxford, 1969, p. 23.
11. G. B. Kistiakowsky and P. H. Kydd, *J. Am. Chem. Soc.* **79**, 4825 (1957).
12. R. T. Meyer, *J. Chem. Phys.* **46**, 967 (1967).
13. R. T. Meyer, *J. Chem. Phys.* **46**, 4146 (1967).
14. R. T. Meyer, *J. Chem. Phys.* **72**, 1583 (1968).
15. D. C. Damoth, *Advances in Analytical Chemistry and Instrumentation,* Vol. 4, John Wiley and Sons, New York, 1956, p. 371; see also this book.

16 D. W. Thomas, *Time-of-Flight Mass Spectrometry*, (Eds. D. Price and J. E. Williams), Pergamon Press, Oxford, 1969, p. 183.
17 R. T. Meyer, *J. Sci. Instr.* **44**, 422 (1967).
18 J. E. Dove and D. McL. Moulton, *Proc. Roy. Soc. (London)* **A283**, 216 (1965).
19 R. T. Meyer, *Time-of-Flight Mass Spectrometry*, (Eds. D. Price and J. E. Williams), Pergamon Press, Oxford, 1969, p.70.
20 B. R. F. Kendall, *J. Sci. Instr.* **39**, 267 (1962).
21 K. A. Lincoln, *Rev. Sci. Instr.* **35**, 1688 (1964).
22 K. A. Lincoln, *Intern. J. Mass Spec. Ion Phys.* **2**, 75 (1969).
23 R. T. Meyer, *Time-of-Flight Mass Spectrometry*, (Eds. D. Price and J. E. Williams), Pergamon Press, Oxford, 1969, p. 66.
24 J. H. Allen, J. F. McKellar, and V. F. Pearce, *J. Sci. Instr.* **40**, 372 (1963).
25 J. M. Freese, and R. T. Meyer, *Rev. Sci. Instr.* **39**, 1764 (1968).
26 R. T. Meyer, C. E. Olson and R. R. Berlint, Sandia Corpn. Report No. SC-R-66-928, August 1966.
27 J. H. Beynon, *Mass Spectrometry,* Elsevier, Amsterdam 1960, p. 319.
28 R. W. Kiser, *Introduction to Mass Spectrometry and its Applications,* Prentice-Hall, New York, 1965.
29 S. C. Barton and J. E. Dove, *Can. J. Chem.* **47**, 521 (1969).
30 A. U. Chaudhry and B. G. Gowenlock, *J. Organometallic Chem.* **16**, 221 (1969).
31 K. Knox, R. G. W. Norrish and G. Porter, *J. Chem. Soc.* 1477 (1952).

Discussion

K. L. Kompa: Supposedly the type of flashtube used will produce a relatively collimated light beam. Shouldn't one then put it away at some distance from the spectrometer to avoid some of the disturbance to the system without losing too much light?

Dr. D. Price: Our flash lamp and reaction vessel design is based on the experience of R. T. Meyer (Sandia Report SC-RR-68-162). It could be that moving the lamp outside the ion source plug box would reduce the electromagnetic disturbance which spoils the first spectrum after the flash. We have not investigated this possibility. However, there would still be interference due to inductance in the lamp firing circuit. This could well outweigh any advantage gained by moving the lamp. It is essential that the firing circuit be designed for minimum inductance.

Dr. G. W. F. Pike: Did you have any long term problems of lead contamination of the mass spectrometer?

Dr. D. Price: We avoid this problem by regular cleaning of the ion source. With the lead tetramethyl partial pressure in the reaction vessel $\lesssim 1$ torr, this is done at weekly intervals. For higher pressures it must be carried out between experiments.

K. L. Kompa: Is there any indication on how much of the volume of the xenon discharge is effective in the photolysis? For instance, does it make a difference if the length of the flash-tube is varied?

Dr. D. Price: We have not made a specific study of this point. The effect of varying the length of the flash-tube would be to vary the time for which the chemical system is exposed to light from the flash. This would affect the amount of light absorbed by the system during the flash.

Dr. J. C. J. Thynne: The formation of ethylene in your reaction system is interesting. Some results of ours, on the reactions of methyl radicals with tetramethylsilane, may cast some light on a possible reaction leading to ethylene formation. When we generated methyl radicals in the gas phase in the presence of $(CH_3)_4Si$ we found small quantities of ethylene to be formed and attribute formation to the reactions:

$$CH_3 + (CH_3)_4Si \rightarrow CH_4 + (CH_3)_3SiCH_2$$

$$\begin{matrix}(CH_3)_3Si-CH_2 \\ (CH_3)_3Si-CH_2\end{matrix} \rightarrow C_2H_4 + (CH_3)_3Si\,Si(CH_3)_3$$

In your system the analogous reactions involving $(CH_3)_4Pb$ might also occur and, since the $(CH_3)_3Pb-Pb(CH_3)_3$ bond is so weak subsequent decomposition to yield $(CH_3)_3Pb$ radicals may well occur.

Chapter 5

An Application of Photoionisation in a Time-of-Flight Mass Spectrometer

P. F. Knewstubb and N. W. Reid

Department of Chemistry, University of Cambridge, England

Photoionisation in a time-of-flight mass spectrometer

The use of photoionisation in mass spectrometry offers several attractive features which have led to its adoption in a large number of cases. For the organic chemist the appeal lies largely in the simplicity of the resulting mass spectra, and also in the fact that the ionising region is cool, so that no complications are introduced by pyrolysis of the sample, or by the temperature dependence of the cracking pattern.

Physical chemists are probably interested in the possibility of ionisation by monochromatic radiation, since a simple control of ionising energy is then possible with no uncertainties due to contact potentials. The energy resolution possible with modern photoionisation equipment is generally much better than that given by electron bombardment sources.

A further advantage is that the ion source can be designed for the efficient extraction of ions, using electric and magnetic fields, without affecting the trajectory of the ionising agent.

Photoionisation is an important modern technique and has been applied to a wide variety of mass spectral problems, but it has never, as far as can be determined from the literature, been used in conjunction with a time-of-flight mass spectrometer.

This paper describes a method which has successfully produced photoionisation mass spectra in a modified Bendix Model 14-101 time-of-flight mass spectrometer. In tackling this new problem the following design criteria were adopted:

1. The light source must deliver flashes of light at a repetition frequency in the region of 8 - 20 kHz, to maintain the full sensitivity of the analogue units.

2. To maintain resolution, the pulse length must be of the order of 0·1 μsec. This appeared to be incompatible with the use of a chopper, so the source itself was to be pulsed.

3. There must be synchronism between the source and the detector circuitry of the mass spectrometer. Since most high voltage switches have comparatively slow response, and are not altogether free of jitter, it appeared that (2) could best be met, by deriving a signal from the lamp to drive the mass spectrometer, rather than *vice versa*.

4. The optical path should preferably be windowless, to avoid the imposition of an artificial upper limit on the ionising energy.

Experimental details

The details of the light source which has been developed for this purpose are indicated in Fig. 1.

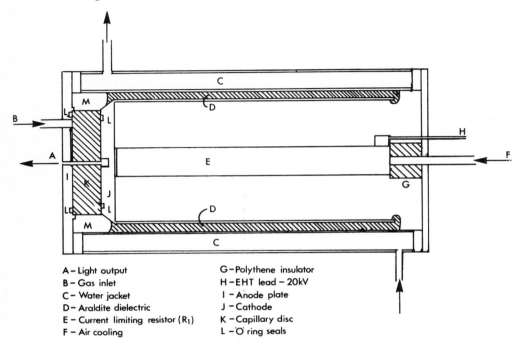

A – Light output
B – Gas inlet
C – Water jacket
D – Araldite dielectric
E – Current limiting resistor (R_1)
F – Air cooling
G – Polythene insulator
H – EHT lead – 20kV
I – Anode plate
J – Cathode
K – Capillary disc
L – O ring seals

Fig. 1: Lamp cross-section.

Following designs of other photoionisation sources, the light is emitted from an intense electrical discharge, constricted by a capillary tube (25 mm long x 0·35 mm bore in boron nitride). The gas pressure (probably a few tenths of a torr) and the nature of the gas (hydrogen and the rare gases) are selected for the optimum performance of the lamp in the wavelength region desired. It is probable that, as in other sources, the light output contains both continuum and line components, which will depend on the conditions of use of the lamp. These features remain to be explored in this apparatus.

The fulfilment of condition (1) of the previous section is achieved by operating the gas discharge as a relaxation oscillator, the capacitor of which is an integral part of the lamp, and having with an 'Araldite' epoxy resin dielectric, a capacity of 1350 pF. The coaxial construction of the capacitor minimises the inductance of the discharge circuit in furtherance of condition (2). The complete circuit is shown in Fig. 2, and the whole of this is contained within the lamp casing to assist with electrical shielding.

Gas is supplied to the lamp via a channel in the anode plate, and escapes through the 0·35 mm light exit hole into the next section of the apparatus. The rate of supply of gas adjusts the gas pressure in the discharge, and this, together with e.h.t. voltage supplied, controls the repetition rate of the discharge, normally set to approximately 10 kHz.

The additional components R_2 and C_2 in Fig. 2 form a potential divider across the discharge circuit, and a 50 V negative pulse appears at their junction every time the lamp

An Application of Photoionisation in a Time-of-Flight Mass Spectrometer

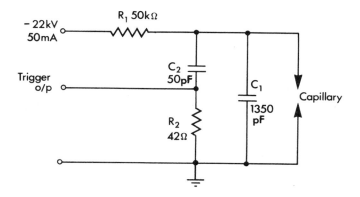

Fig. 2: Electrical circuit of the lamp.

flashes. This is used to trigger a Solartron Model OPS 100 C pulse generator, whose output drives the mass spectrometer. The mass spectrometer master clock is disconnected.

It is possible to adjust the values of R_2 and C_2 to produce a larger pulse, which drives the mass spectrometer directly. However, it is of very short duration, so rather a large amplitude is required to give reliable triggering. This was found to cause damage to the semiconductors in the Bendix circuitry, accordingly the pulse generator, a valve model, was interposed as a buffer stage.

Typical performance of the lamp is shown in Fig. 3. This is the output of a 1P28 photomultiplier sensitised by sodium salicylate as displayed on a Tektronix Model 543 oscilloscope, with a rise time of about 12 nsec. The upper portion shows several flashes on a time scale of 50 µsec/cm. Several traces are superimposed, so the slight broadening

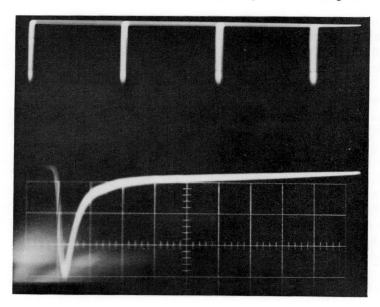

Fig. 3: Oscilloscope trace showing lamp output. Time scale: upper trace 50 µsec/cm, lower trace 0·1 µsec/cm.

of the peaks towards the right reflects the uncertainty in the repetition rate. The lower trace, at 0·1 μsec/cm, shows a single flash, and it can be seen that there is very little light output after about 130 nsec.

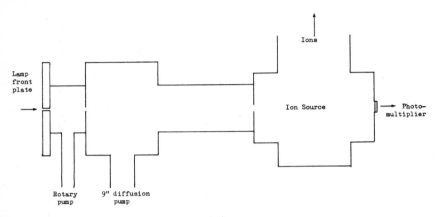

Fig. 4: Schematic diagram of the optical coupling.

The optical coupling is shown in Fig. 4, and is by direct line of sight into the ion source. The light source is differentially pumped in two stages, and typical pressures are 0·1 to 1, and 10^{-5} torr, respectively.

The source as described could be put to use with a standard Bendix ion source, with appropriate defining slits on the beam axis. The further modifications to be described are a second part of this work, which is directed towards observations of ions decomposing with fairly long life times — so-called metastable ions.

Figure 5 shows a normal photoionisation mass spectrum (i.e. no metastable transitions). The sample is benzonitrile, and the lamp gas is hydrogen. Only the parent ($m/e = 103$) and the daughter ($m/e = 76$) peaks are found. The sharp spikes on the trace are spurious, as has been determined by comparing several scans of the same spectrum.

The observation of 'metastable ions'

In normal operation of the time-of-flight instrument, no effects attributable to 'metastable ions' are observed. The decomposition of an ion into two or more lighter fragments occurs with negligible change of velocity or of flight time, and hence ions seen as fragments must have reached that state in the ion source before being launched down the flight tube. Metastable ions will appear in the position of the parent peak.

Observations of metastable ions in a T.O.F. instrument have previously been made by a modification described by Hunt *et al.*[1] This involves the erection of a partial potential barrier in the flight tube. The method has been further adapted by Ottinger,[2] who introduced the idea of allowing ions to decompose in a potential gradient. This principle is given full rein in the present work.

The effect of the modification is to produce a uniform axial electric field over the whole flight path. For an ion decomposition reaction.

$$m_1^+ \rightarrow m_2^+ + n$$

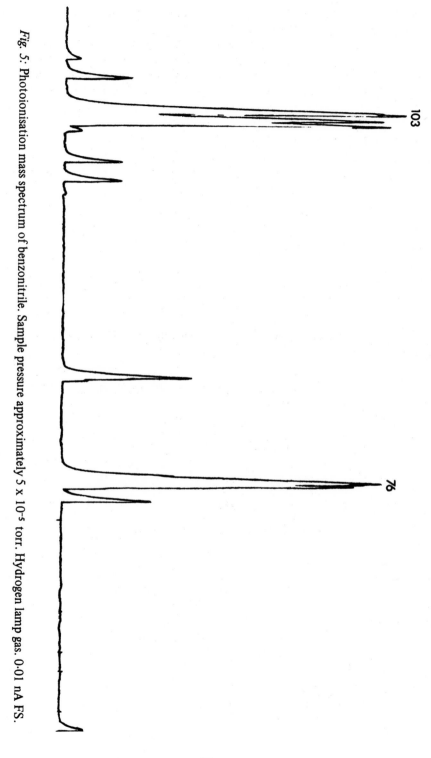

Fig. 5: Photoionisation mass spectrum of benzonitrile. Sample pressure approximately 5 × 10⁻⁵ torr. Hydrogen lamp gas. 0·01 nA FS.

any parent ions which traverse the flight path without decomposing give rise to a peak at twice the normal flight time. Similarly, if the decomposition takes place within about 0·1 μsec a daughter peak will be found, likewise at double the normal flight time.

If, however, the parent ion decomposes in flight, the charged fragment will arrive at the detector with a total flight time intermediate between the two values found above, and related to the time of decomposition, as shown by the following analysis of the motion:

Table 1. Notation for equation of motion of ions

	Instant of ionisation	Instant of decomposition	At detector cathode
Time	0	T	$T + t$
Velocity	0	V_1	V_2
Distance covered	0	x	d

Referring to the notation of Table 1, the following relations are easily derived, if the field strength is E V/m:

$$V_1 = \frac{eET}{m_1} \quad \text{where} \quad e = \text{electronic charge}$$

$$X = \frac{eET^2}{2m_1}$$

$$V_2 = V_1 + \frac{eEt}{m_2}$$

$$d - x = V_1 t + \frac{eEt}{2m_2}$$

$$= \frac{eE}{m_1} tT + \frac{eE}{2m_2} t^2$$

$$t^2 + \frac{2m_2}{m_1} Tt - (d-x)\frac{2m_2}{eE} = 0$$

$$\therefore \quad t = -\frac{m_2}{m_1}T + \left[(m_2/m_1)^2 T^2 + 2(d - \frac{eE}{2m_1}T^2)\frac{m_2}{eE}\right]^{1/2} \quad (1)$$

Total flight time, $T_f = t + T$

In fact this analysis is slightly simplified, since a further term must be introduced to allow for the flight time down the field-free region provided by the multiplier stack. This term is

$$t^l = l/V_2$$

where l is the length of the multiplier stack. This term is not necessary for a qualitative understanding of the operation of the modified mass spectrometer, and it may be neglected at this stage. It has been included in all quantitative work.

The result is that any 'metastable ions' decomposing down the flight tube will give rise to 'smeared out' parent and daughter peaks, whose shape can be further analysed, as below, to yield the decomposition function.

The ion current (I) at a decomposition time, t, (transformed from flight time via Eqn. (1)) is proportional to the number of decompositions taking place between t and $t + \Delta t$, where Δt is the gate width, also transformed to a scale of decomposition time.

i.e. $\quad I \alpha (e^{-kt} - e^{-k(t + \Delta t)})$ for a single rate coefficient

$$\therefore \quad \log I = \text{Const.} + \log(1 - e^{-k\Delta t}) - kt \quad (2)$$

Since Δt is only a slowly varying function of t, a semilog plot of ion current versus decomposition time will give a straight line, whose slope yields the rate coefficient directly.

In the more probable case, viz. that where a distribution of rate coefficients is operative, the semilog plot will be a curve, and the analysis is more complicated.

An implicit assumption in the above treatment is that the multiplier gain is independent of ion velocity. This is clearly not so, but for all cases so far treated, a calculated velocity dependent correction to the ion current proves, in fact to be negligible.

Modifications to the mass spectrometer

The ion lens, x and y deflection plates, and drift tube liner have all been removed, and replaced by a series of carefully designed plates (Fig. 6). These are fed from a potential divider, with the e.h.t. applied to the multiplier end, so that the flight tube is now a region of constant (about 30 V/cm) rather than zero field.

Figure 7 is a resistive paper plot of the equipotentials inside a section of this flight tube. It shows very clearly the cross section of the plates. The shape was chosen so that the constant field region is shielded from the perturbing effects of the earthed walls of the flight tube.

The equipotentials are evenly spaced, confirming the constancy of the field, and their slight curvature is interesting, since it predicts a certain focusing effect, which should reduce the loss of ions having a radial component of velocity.

The optical slits are machined in the walls of the first of these plates, so that the

Fig. 6: View of the modified flight tube from the source end. The second plate has a smaller central hole than all the others in an attempt to raise the source pressure.

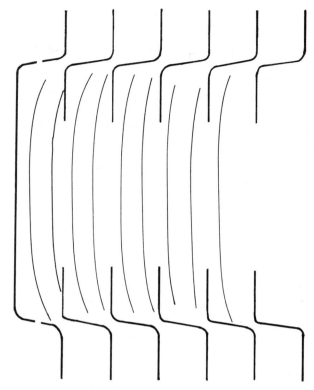

Fig. 7: Resistive paper plot of the field inside the modified flight tube.

photoions are formed in a region of constant field. This immediately poses the question of whether the photoionisation mass spectra obtained here are genuine or whether they are false ones caused by the impact on the sample of photoelectrons accelerated in the ion source field.

Apart from the simplicity of the spectra, which is what one expects of photoionisation, there are two reasons for believing these to be genuine.
(i) If one considers the mass spectrum of toluene, the peaks at m/e = 91 and 92 will be separated by less than a millimeter when they reach the multiplier cathode. Thus, if the ionising region extends for more than about a millimetre along the axis of the flight tube the resolution of the instrument will be seriously degraded.

An experiment was performed with the constant field extending about 2·5 cm 'downstream' from the point of ionisation. Any photoelectrons released would thus have had sufficient energy to cause ionisation of the sample over at least 2 cm of their travel, but no impairment in resolution was observed.
(ii) Again using a toluene sample, the mass spectrum was recorded with field strengths in the range 5-30 V/cm. If photoelectrons had been responsible for the ionisation, this procedure would effectively have scanned over the breakdown diagram of the sample, giving an alteration in the various relative peak heights. Such a variation was not found.

In view of the above results it was concluded that genuine photoionisation mass spectra have been obtained.

Use of the modified mass spectrometer

As a test of the apparatus the previously reported[3] metastable transition

$$(CH_3O)_3 PO^+ \rightarrow (CH_3O)_2 HPO^+ + [HCHO]$$
$$m/e \qquad\qquad 140 \qquad\qquad\qquad 110$$

seen in the mass spectrum of trimethyl phosphate was studied.

Figure 8 shows a scan of the mass range of interest on this occasion. This particular scan was taken with 15 V electron impact at a sensitivity of 0·01 nA at full scale, and is reproduced here since it contains no range changes and thus gives a better idea of the peak shape found.

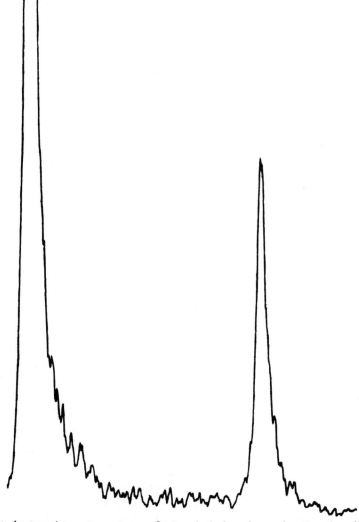

Fig. 8: 15 volt electron impact spectrum of trimethyl phosphate, showing the effect of the metastable transition $(CH_3O)_3 PO^+ \rightarrow (CH_3O)_2 HPO^+ + [HCHO]$. Scan rate 2. Time constant 1 sec.

The spectra which were recorded and used in a quantative analysis, according to Eqn. (1) and (2) were obtained with 50 V electrons, and thus at a high sensitivity. The portion between the two peaks, i.e. the ion current due to decompositions in the flight tube, was normally scanned at maximum possible gain, to improve the accuracy of the measurement.

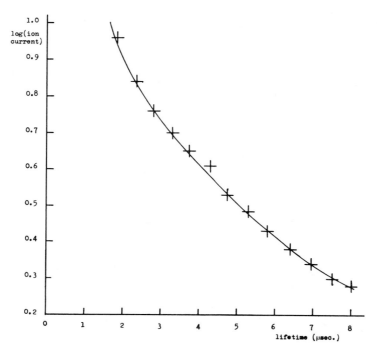

Fig. 9: Plot of $\log_{10}$ (ion current) versus decomposition time for the trimethyl phosphate metastable transition. $(CH_3O)_3PO^+ \rightarrow (CH_3O)_2HPO^+ + (CH_2O)$.

Figure 9 shows the semilog plot obtained by averaging several scans. It is a smooth curve, indicating that there is more than one rate coefficient for the decay. Since the curve does not become a straight line at the longest accessible times it appears most likely that a distribution of rate coefficients is operative, in agreement with the predictions of the quasi-equilibrium theory.

Summary

A photoionisation facility has been added to a Bendix time-of-flight mass spectrometer. The instrument has been further modified to make it capable of measuring ion lifetimes in the range of microseconds. One further modification is planned — a Seya-Namioka monochromator has been constructed and will be installed to allow a study of the decompositions of the ions formed by photons of known energy.

Acknowledgements

We wish to acknowledge, with thanks, the support of one of us (N.W.R.) by maintenance grants from the Elsie Ballot Scholarship and the South African Council for

Scientific and Industrial Research. Thanks are also due to the Royal Society for a grant for apparatus.

References
1 W. W. Hunt, R. E. Hoffman and K. E. McGee, *Rev. Sci. Instr.* **35**, 82 (1964).
2 C. H. Ottinger, *Z. Naturforsch.* **22a**, 20 (1967).
3 D. L. Dugger and R. W. Kiser, *J. Chem. Phys.* **47**, 5054 (1967).

Discussion
Dr. G. W. F. Pike: Other sources of flight time change, other than metastable transitions that come to mind are:
    1 ion-molecule collisions,
    2 metastable state energy changes,
    3 initial ion source dispersion,
    4 kinetic energy of fragmentation.
Were any of these processes considered significant in your work?

N. W. Reid: Ion-molecule reactions are certainly capable of causing a change in flight time, but it is easy to test for these, and, in fact, any other collision-dependent effect, by noting the pressure dependence of the results.

The effect of kinetic energy of decomposition was studied by means of a computer simulation. This showed that changes are caused only in the immediate vicinity of the parent and daughter peaks, i.e. flight time changes are small, corresponding to one or two mass units on the recorded spectrum. On the other hand, changes in flight time caused by metastable transitions cover the whole region between parent and daughter flight times.

Dr. P. F. Knewstubb: The computer calculation analysing the possible flight times of ions tested not only the effect of kinetic energy release in the decomposition, but also that of thermal velocities of molecules in the source and of optical beam width. None of these produced very great effects on the predicted curve of arrival time, so that for the present at least we feel that these effects can be ignored.

Chapter 6

Photoelectron Spectrometry and R.P.D. Measurements on Sulphur Hexafluoride

Jacques Delwiche*

Université de Liège, Belgium

Introduction

In recent years, a considerable improvement in knowledge of the ionised gases has been obtained by the use of the retarded potential difference (R.P.D.)[1] method of measuring the ionisation efficiency curves. Usually, the ion source is coupled with a mass spectrometer to perform the mass analysis of the ions created.

The aim of this chapter is to describe, first, the adaptation of a five-slit electron gun on a Bendix time-of-flight mass spectrometer (Model 14) and then to show how the use of the R.P.D. technique combined with photoelectron spectrometry[2,3] allows the study of the behaviour of the molecules under the action of ionising particles.

T.O.F. mass spectrometers are especially well suited for the use of the R.P.D. technique for various reasons:

(a) ionisation takes place in a field-free region as the spectrometer is pulsed;
(b) the ion source is operated near ground potential so that insulation problems are simplified;
(c) the source magnets are located outside the vacuum envelope; this simplifies the collimation of the electron beam;
(d) constant focussing of a particular mass is made easy by the absence of an analysing magnet.

The setting up of the electron gun

Several modifications have been introduced in the electronics and in the electron and ion optics of the spectrometer in order to increase its stability and its sensitivity.
Modifications of the electronics. In the early stages of development, it appeared that almost all the pulses of the mass spectrometer were varying very slowly in amplitude. The gate pulses were also oscillating around their normal position in the mass scale. These difficulties were overcomed by using storage batteries for the heating current of the filaments of the electronic valves in chassis 2 and 3 and in the two analogue output systems. Later, the batteries were replaced by regulated d.c. power supplies.

---

* 'Charge de Recherches' of the F N R S of Belgium

The control of the accelerating voltage and of the various d.c. bias of the five-slits electron gun. The control of the electron energy as provided with the mass spectrometer was too coarse to permit the measurement of ionisation efficiency curves. It was replaced by the circuit presented in Fig. 1. This circuit also provides the potentials for the five slits of the electron gun. Power is supplied by three mercury cells.

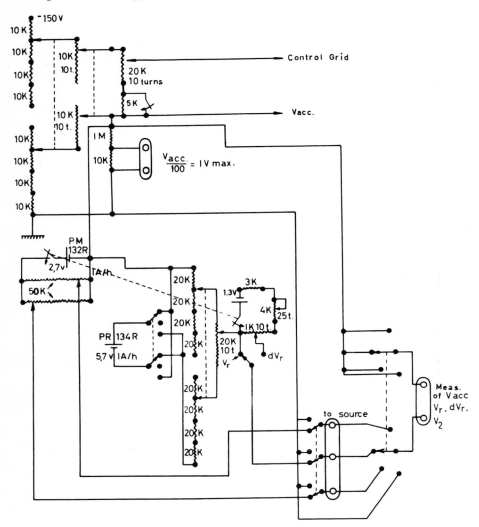

Fig. 1: Circuitry giving the different d.c. bias of the ion source.

The polarity of slit 3 can be changed. This has proved to be useful during the adjustments of the electron gun. Separate controls are provided for slits 2 and 4. Experience has shown that this is not necessary. Both slits are now internally connected and the adjustment of their potential is carried out with the same potentiometer. The best results are obtained when these slits are slightly positive. Their potential, however, does not affect the shape of the electrons' energy distribution (measured by varying the potential of slit 3). This is shown by Figs. 2 and 3. They show the variation of the trap current with the retarding potential for different potentials on slits 2 and 4.

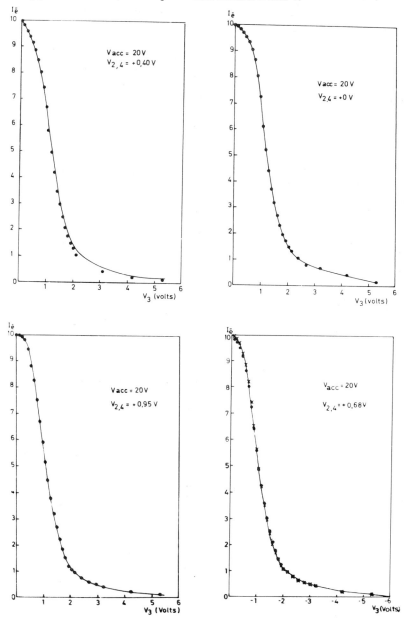

Fig. 2: Variation of the trap current (in arbitrary units) with the retarding potential (V_3). ● without H.T., x with H.T.

In the normal Bendix ion source, the trap is operated at +150 V. Ions created in the vicinity of the trap can be accelerated back into the ionisation region,[4,5] collected and focussed as normal ions. This gives measurable ion currents well below the normal ionisation energies. Therefore, the trap potential was reduced to 47·25 V. A further decrease of the trap potential was found to be unnecessary. The electron current was recorded using a Leeds & Northrup micromicroampere indicating amplifier.

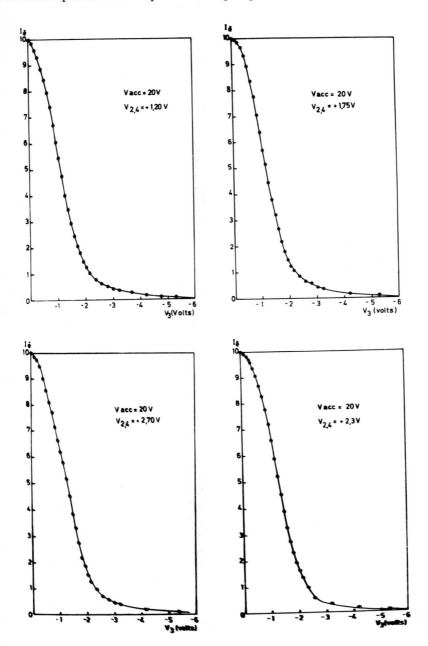

Fig. 3: Variation of the trap current (in arbitrary units) with the retarding potential (V_3).

The trap current regulation

The R.P.D. method of measurements precludes the use of a trap current stabiliser. One must, therefore, carefully stabilise the heating current of the filament. This was accomplished by using high capacity storage batteries (6 V and 152 amp hours) and the transistorised device shown in Fig. 4. The short time variations of the trap current are less than 1% and the temperature compensation (8 Ω thermistor) gives excellent long term stability.

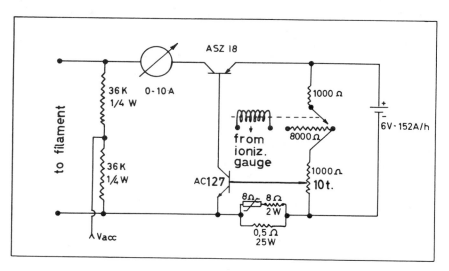

Fig. 4: Transistorised current stabiliser for the filament.

A relay connected to the vacuum gauge automatically reduces the heating current to a safe value if the pressure becomes too high in the flight tube.

The modifications to the ion source. Rhenium filaments have been used since experience has shown that they give a better short term stability than tungsten filaments. They are less sensitive to the nature of the substance studied. Unfortunately, their evaporation rate is high and therefore the electron gun needs frequent cleaning.

The electron gun has been modified by the addition of a small square plate behind the filament. Experience showed that the best results were obtained with this plate connected to the positive side of the filament. Its presence dramatically diminished the influence of the source magnet's position on the characteristics of the ion source.

Several changes were made in order to increase the pressure in the ionisation region. First, a Teflon diaphragm was placed between the flight tube and the last accelerating grid of the ion source. Its aperture was 6 x 13 mm. It resulted in a definite increase in sensitivity, but the high pressure in the vicinity of the filament caused a rapid dirtying of the electron gun. The pumping diaphragm was removed and replaced by a closed ionisation chamber. Entrance and exit electron slits were 2 x 5 mm wide. The ion exit slit was 6 x 6 mm wide. The backing plate was removed and the ionisation chamber was insulated from the source support by a thin Mylar ribbon. The best focussing of the ions was obtained by connecting the ionisation chamber to the ion focus grid. The resulting increase of the capacity between the ion focus grid and the source support does not affect the rise time of the ion focus pulse. All the surfaces of the electron gun and of the con-

nection chamber were coated with Aquadag.
Adjustment of the electron gun. Each time that the ion source was removed from the mass spectrometer, some adjustments were necessary to obtain a good collimation of the electron beam. These will now be described.

First, the source magnets are displaced until one gets a trap current of 1×10^{-6} A for a heating current of about 2·5 - 2·9 A, the retarding potential being zero. This is done with the high voltage applied on the accelerating grids and in presence of argon. One must then verify that the trap current does not vary more than 5% when the accelerating voltage is changed from 5 V to 40 V. The variation must be linear. The d.c. bias of the control grid (slit 1 of the electron gun) is then adjusted to obtain the first maximum of trap current (about 8 V of extracting potential). Figure 5 presents the variation of the trap current

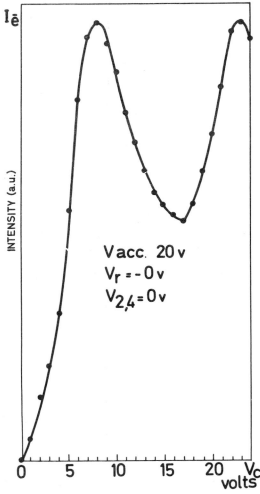

Fig. 5: Variation of the trap current with the potential of the control grid (slit 1).

with the potential of the control grid. The potential of slit 3 (V_r) is then adjusted to operate the electron gun in the middle of the energy distribution of the electrons (usually around

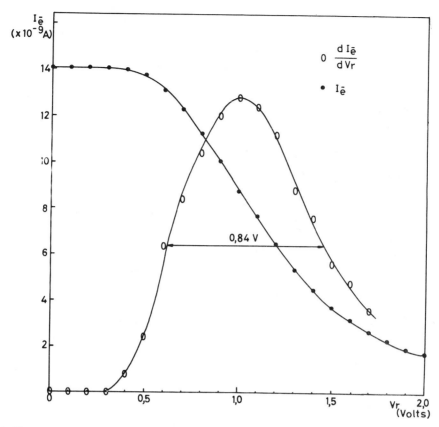

Fig. 6: Variation of the trap current with the retarding potential (V_r).

— 1·0 V as shown in Fig. 6). The variation of trap current, for 20 eV electrons, must be at least 15% for a 0·1 V change of the retarding potential. It has sometimes been observed that when the source magnets are not well adjusted the trap current increases when the retarding potential is made more negative. The position of the source magnets must then be changed and all the adjustments described above must be made once again.

The next step is to verify the proportionality between the trap current and the ion current. One must then ascertain that the ion current for argon is no longer measureable 0·5 eV below its ionisation energy and that below this energy the ion current does not vary for a small change of the retarding potential. It may happen that the ion current increases while the trap current decreases, when the retarding potential is made more negative. One must then change the position of the source magnets.

The measurement of the ionisation efficiency curves. For normal working a trap current of 0.5×10^{-6} A was used. For ions of very small intensity, this could be increased up to 2×10^{-6} A. The pressure in the ionisation chamber was adjusted in order to have, for the parent ion, an electron current at the output of the multiplier of about 1×10^{-8} A with ionising electrons having an energy 5 eV above the ionisation energy of the molecule.

For the ionisation energies of parent ions, a rare gas was used to provide an energy standard. Both ion currents were recorded simultaneously on a three-pen recorder (Rikadenki B-34). The third pen was used to record the trap current. The ionisation energy of the parent ion was used as a standard to determine the ionisation energy of the fragment ions.

Results and discussion

Neon and argon. The ion source has been tested using rare gases, as these are often used as standards for the determination of ionisation energies. Results obtained for neon and argon are briefly reported.

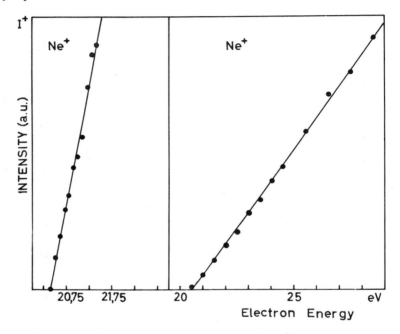

Fig. 7: The ionisation efficiency curve (R.P.D. method) of Ne^+. The energy scale is corrected by using the spectroscopic value of the ionisation energy.

Figure 7 presents the R.P.D. curve for neon. The absence of structure is in agreement with the results of photoionisation[6] and of R.P.D. measurements made by Fox.[7] The linearity of the curve seems to indicate that the space charge effects and the effects of relaxation of energy described by Marmet[8] are negligible.

For argon, our results together with those of other workers using different techniques are summarised in Table 1.
As it can be seen, our results are consistent with the literature data.

Sulphur hexafluoride. The intensity of the parent ion SF_6^+ in the mass spectrum of SF_6 is extremely low: the relative intensities of $SF_6^+/SF_5^+/SF_4^+/SF_3^+$ are respectively: $3 \times 10^{-4}/1 \cdot 0/0 \cdot 11/0 \cdot 14$ at $21 \cdot 21$ eV.[23] This very low intensity precludes the study of the parent ion by the R.P.D. technique. One must therefore obtain information about the position of the electronic states of SF_6^+ by another method of investigation. Photoelectron spectrometry is well suited for this purpose as it allows the study of electronic energy levels that are unstable or that have a very short lifetime.

Table 1. Energy levels of Ar⁺ (in eV above threshold)

$^2P_{1/2}$										Method	Reference
0.17	0.34	0.62	–	0.91	–	–	–	–	–	Selector	1
0.21	–	0.64	–	–	1.27	–	–	–	–	Selector	10
0.19	–	0.58	0.77	0.97	1.30	–	1.75	–	–	Selector	11
0.20	–	0.50	0.78	–	1.27	–	1.74	–	2.45	Selector	12
0.17	–	–	–	–	–	–	–	–	–	Photoionisation	13
0.21	–	–	–	–	–	–	–	–	–	Photoionisation	14
0.20	0.39	0.57	–	0.90	1.33	1.55	1.70	1.90	2.15	Photoionisation	15
0.17	–	–	–	–	–	–	–	–	–	Photoionisation	16
0.178	–	–	–	–	–	–	–	–	–	Photoelectrons	3
0.20	–	–	–	–	–	–	–	–	–	R. P. D.	17
0.20	–	–	–	–	–	–	–	–	–	R. P. D.	18
0.20	–	0.55	–	1.0^a	–	–	–	–	–	R. P. D.	19
0.20	–	0.55	–	1.0^a	–	–	–	–	–	R. P. D.	20
0.20	curvature on 1.6 eV									R. P. D.	21
0.18	0.42	–	–	1.01	1.36	–	–	1.96	–	Winters	22
0.20	–	0.55	–	1.0	–	–	–	2.10	–	R. P. D.	This Work

^{a} The results were not published for higher energies.

The photoelectron spectrum represented in Fig. 8 was taken on a magnetic deflection analyser devised by May.[24] We used the 584 Å resonance line of helium. Argon was used to calibrate the energy scale. The high resolution spectrum (Fig. 9) was obtained on an electrostatic 127° energy analyser recently described by Turner.[25]*

The most probable transition to the fundamental state of the ion SF_6^+ takes place at 15.81 eV, but the shape of the corresponding photoelectron band shows that the adiabatic ionisation energy must be located at a lower energy. We estimate it to be 15.30 eV. This value is in good agreement with that of 15.35 eV obtained by Frost and McDowell[26] with a spherical grid analyser.

The second electronic energy level of SF_6^+ is located at 16.70 eV; the most probable transition occurs at 17.10 eV. Lui and co-workers[27] and Nostrand and Duncan[28] have observed in the absorption spectrum of SF_6 four bands, three of which can form a Rydberg series converging at 16.15 eV. We do not observe any structure at this energy in our photoelectron spectrum. It is possible, however, that the cross section of that energy level is too small to allow its detection with 21.21 eV photons. However there is also no corresponding structure in the ionisation efficiency curves obtained by photon impact.

---

* We are greatly indebted to Dr. D. W. Turner for the communication of this high resolution spectrum.

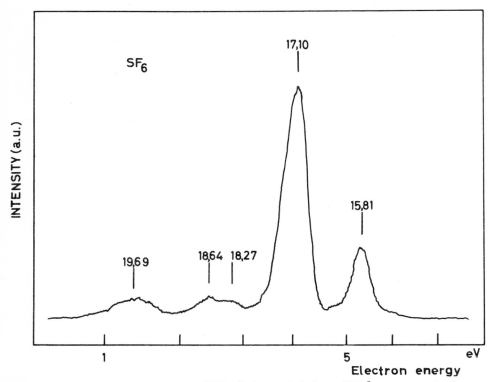

Fig. 8: The photoelectron spectrum of SF_6 (light used: helium 584 Å resonance line).

Therefore, the existence of an electronic state of SF_6^+ at 16·15 eV cannot be confirmed by our measurements. The other electronic bands have their maximum at 18·27 eV, 18·64 eV and 19·30 eV. These values correspond to vertical transitions. The adiabatic transitions take place at 18·10 eV, 18·50 eV and 19·30 eV. Two of these values agree with those of Frost et al.[26] (18·11 eV and 19·50 eV). These authors have not observed an energy level at 18·50 eV. This may be due to a too low resolution power of their analyser.

The last electronic band of the spectrum shows a vibrational structure that stops with the ninth level. There is no decrease of the energy difference between the vibrational levels (see Table 2). As we observe a break in the R.P.D. curve of SF_5^+ at 19·35 eV, it seems that the SF_6^+ ion is predissociated and that this predissociation is extremely rapid.

The shape of our R.P.D. ionisation efficiency curve for SF_5^+ (Figs. 10 and 11) is similar to the curve obtained by Fox and Curran [33] without mass analysis. Our value of the appearance energy (15·75 ± 0·05 eV) agrees very well with the value of 15·9 eV determined by Dibeler and Mohler.[29]

The appearance energy of SF_5^+ and the first break on the R.P.D. curve (16·90 eV) coincide with the position of the two first bands in the photoelectron spectrum of SF_6^+. This may indicate that the Franck-Condon zone crosses the energy surface of these two energy levels above one of their dissociation asymptotes.

As shown by Ahearn and Hannay,[30] an appearance energy of 16 eV for F^- suggests the existence of a second mechanism for production of the SF_5^+ ions by decomposition of

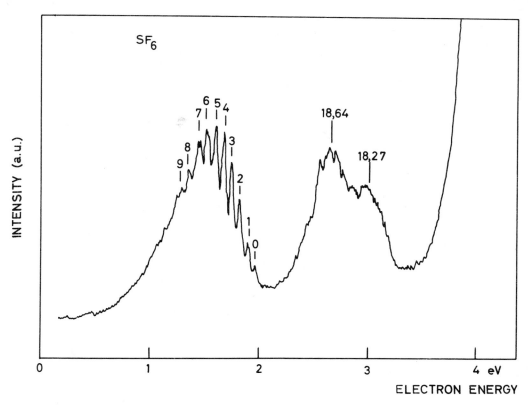

Fig. 9: Part of the photoelectron spectrum of SF_6 obtained with a high resolution 127° electrostatic energy analyser (light used: helium 584 Å resonance line).

Table 2. Energy levels of SF_6^+; SF_5^+; SF_4^+; SF_3^+ (in eV)

SF_6^{+a}	Reference 33	This Work SF_5^{+b}	SF_4^{+b}	SF_3^{+b}
15.81	15.85 ± 0.15	15.75 ± 0.05	18.50 ± 0.10	19.80 ± 0.10
17.10	17.00 ± 0.20	16.90 ± 0.10		
18.27	18.00 ± 0.20	17.90 ± 0.10	21.75 ± 0.10	21.50 ± 0.10
18.64		19.35 ± 0.10		
19.69[c]		20.35 ± 0.10	24.10 ± 0.20	22.50 ± 0.10
		21.65 ± 0.10		
		23.65 ± 0.10		24.20 ± 0.10
		26.80 ± 0.10		

[a] Photoelectron spectrum
[b] R.P.D. curves

c   Vibrational levels: $v'' = 0 : 19.32$ eV; $v'' = 1 : 19.40$ eV; $v'' = 2 : 19.47$ eV;
$v'' = 3 : 19.54$ eV; $v'' = 4 : 19.62$ eV; $v'' = 5 : 19.69$ eV;
$v'' = 6 : 19.75$ eV; $v'' = 7 : 19.83$ eV; $v'' = 8 : 19.91$ eV;
$v'' = 9 : 19.99$ eV.

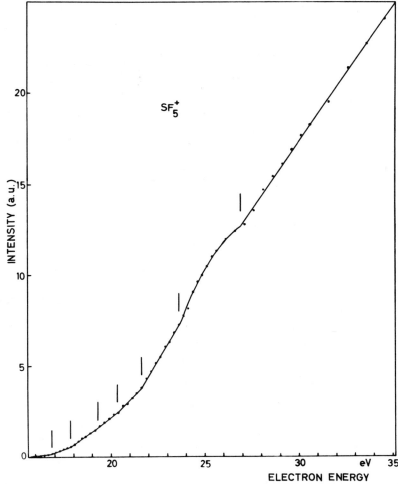

Fig. 10: The ionisation efficiency curve (R.P.D. method of SF_5^+). The breaks are indicated by vertical lines. Reference: argon.

a super-excited neutral molecule (ion-pair process). This is confirmed by the presence of a maximum in the photoionisation curve of SF_5^+ at 16 eV. The cross section for that mechanism is too low to permit its detection in our R.P.D. curves.

The photoionisation curves of SF_5^+ [23] shows that the break located at 17.90 eV in the R.P.D. curve must be attributed to autoionisation processes.

Codling[31] in his study of the absorption spectrum of SF_6 found a vibrational structure between 20.83 eV and 22.29 eV. It is divided into two parts; the first ranges from 20.83 eV to 21.11 eV and the second from 21.56 eV to 22.29 eV. The energy

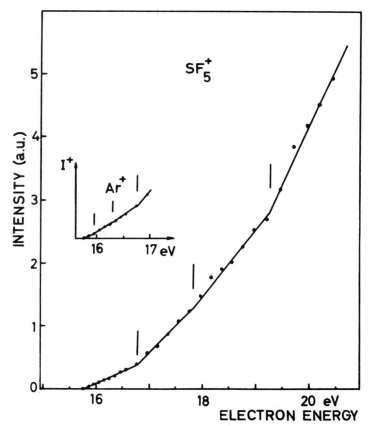

*Fig.11:*Detail of the ionisation efficiency curve (R.P.D. method) of SF_5^+ near the appearance energy. Reference: argon. The breaks are indicated by vertical lines Upper curve: ionisation efficiency curve of Ar^+.

difference between the vibrational levels is the same in both parts. The author could not decide whether these vibrational structures belong to the same electronic energy level or to two different levels. We observe that these two vibrational structures start at energies close to those of two breaks present in our R.P.D. curve of SF_5^+: 20·35 eV and 21·65 eV. It seems therefore that we have two different electronic states, both being pre-ionised and decomposing into $SF_5^+ + F + e^-$.

Between 23·65 eV and 26·80 eV our R.P.D. curve is not linear. This curved structure coincides with the existence of a Rydberg series converging at 26·83 eV.[31] We conclude that the terms of this series are pre-ionised and that their autoionisation is dissociative. The break at 26·80 eV shows the appearance of a new excited state of SF_6^+, leading to the formation of SF_5^+ by predissociation.

Our experimental value for the appearance energy of SF_4^+, 18·5 ± 0·1 eV, agrees with the value of 18·9 eV obtained by Dibeler and Mohler.[29] This energy coincides with the existence of an electronic band in the photoelectron spectrum of SF_6: 18·5-18·64 eV. We conclude that the SF_4^+ ions are created from excited and unstable SF_6^+ ions.

The first break (see Fig. 12) located 3·25 eV above the appearance energy of the

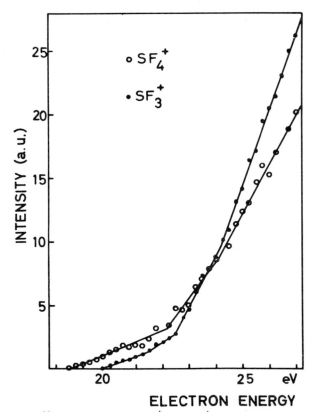

Fig. 12: The ionisation efficiency curves of SF_4^+ and SF_3^+. Reference: appearance energy of SF_5^+.

ion, cannot be associated with the simultaneous formation of two fluorine atoms and a SF_4^+ ion as the dissociation energy of F_2 is 1·634 eV.[32] We attribute this break to the appearance of a new electronic state of SF_4^+.

For the same reason, the break situated at 24·1 eV must also correspond to an excited electronic state of SF_4^+.

The appearance energy of 20·1 eV for SF_3^+ measured by Dibeler and Mohler[29] agrees with our value of 19·80 eV $\pm$ 0·1 eV.

The first break (Fig. 12) is situated 1·7 eV above the threshold. This value is very close to the value of 1·634 eV for the dissociation energy of fluorine.[32] It seems, therefore, that at the appearance energy of the ion SF_3^+ is formed together with a fluorine molecule while at 21·5 eV, SF_3^+ is formed together with three fluorine atoms.

The two next breaks are also separated by 1·7 eV. We conclude that this is due to the formation of an excited state of SF_3^+ with the formation of F_2 + F at 22·5 eV and of 3 F at 24·2 eV.

References

1   R. E. Fox, W. M. Hickam, D. J. Groove and T. Kjeldaas Jr., *Rev. Sci. Instr.* **12**, 1101 (1955).

2 B. L. Kurbatov, F. I. Vilesov and A. N. Terenin, *Dokl. Akad. Nauk SSSR* **138**, 1329 (1961); **140**, 797 (1961).
3 D. W. Turner, *Nature* **213**, 795 (1967).
4 J. Momigny, *Bull. Soc. Chim. Belges* **66**, 33 (1957).
5 V. Cermak and Z. Herman, *Nucleonics* **19**, 106 (1961).
6 PoLee, G. L. Weissler, *Proc. Roy. Soc. (London)* **A219**, 71 (1953).
7 R. E. Fox, *J. Chem. Phys.* **35**, 1379 (1959).
8 P. Marmet, *Can. J. Phys.* **42**, 2120 (1964).
9 C. Hutchison, *Advances in Mass spectrometry,* Institute of Petroleum, London, 1961.
10 S. N. Foner and B. H. Nall, *Phys. Rev.* **122**, 512 (1961).
11 L. Kerwin, P. Marmet and E. Clarke, *Advances in Mass Spectrometry,* Institute of Petroleum, London, 1961.
12 C. E. Brion, D. C. Frost and C. A. McDowell, *J. Chem. Phys.* **44**, 1034 (1966).
13 R. E. Huffman, Y. Tanaka and J. C. Larrabee, *J. Chem. Phys.* **39**, 902 (1963).
14 E. Schonheit, *Z. Naturforsch.* **16a**, 1094 (1961).
15 F. J. Comes and W. Lessman. *Z. Naturforsch.* **16a**, 1396 (1961).
16 V. H. Dibeler et al., *Advances in Mass Spectrometry,* Institute of Petroleum/ASTM, London, 1964.
17 J. E. Collin, *Advances in Mass Spectrometry,* Institute of Petroleum, London, 1959.
18 R. E. Fox, W. M. Hickam and T. Kjeldaas Jr., D. J. Groove, *Phys, Rev.* **84**, 859 (1951).
19 H. Sjogren and E. Lindholm, *Phys. Letters* **4**, 85 (1963).
20 C. E. Melton and W. A. Hamill, *Notre Dame Report,* 1963.
21 Y. Kaneko, *J. Phys. Soc. Japan* **16**, 1587 (1961).
22 R. E. Winters, J. H. Collins and W. L. Courchenne, *J. Chem. Phys.* **45**, 1931 (1966).
23 V. H. Dibeler and J. A. Walker, *J. Chem. Phys.* **44**, 4405 (1966).
24 D. W. Turner and D. P. May, *J. Chem. Phys,* **45**, 471 (1966).
25 D. W. Turner, *Proc. Roy. Soc. (London),* **307A**, 15 (1968).
26 D. C. Frost, C. A. McDowell, J. A. Sandhu and D. A. Vroom, *J. Chem. Phys.* **46**, 2008 (1967).
27 T. K. Lui, G. Moe and A. B. F. Duncan, *J. Chem. Phys.* **19**, 71 (1951).
28 E. D. Nostrand and A. B. F. Duncan, *J. Am. Chem. Soc.* **76**, 3377 (1954).
29 V. H. Dibeler and F. L. Mohler, *J. Res. Natl. Bur. Stds.* **40**, 25 (1948).
30 A. J. Ahearn and N. B. Hannay, *J. Chem. Phys.* **21**, 119 (1953).
31 K. Codling, *J. Chem. Phys.* **44**, 4401 (1966).
32 S. W. Benson, *J. Chem. Educ.* **42**, 502 (1965).
33 R. E. Fox and R. K. Curran, *J. Chem. Phys.* **34**, 1595 (1961).

Chapter 7

The Use of an Ion Probe Technique for Investigating Surface Reactions: The Synthesis of Deutero-Ammonia on Pure Iron

J.C. Robb, D. R. Terrell and D. W. Thomas

Chemistry Department, University of Birmingham, England

Introduction

In the last decade, several new physical methods of investigating phenomena associated with adsorbed layers on solids have been developed. They include low energy electron diffraction, electron- and photon-impact desorption, electron spectroscopy for chemical analysis (ESCA), field ionisation and the field atom probe and ion-impact desorption. Ion impact has frequently been used in the past merely to clean surfaces of adsorbents and where mass analysis was available, hydrocarbon ions characteristic of the pump oils were observed. Ion impact is not totally destructive in the sense that molecular species are sputtered to yield only atomic ions or very low molecular weight fragment ions, e.g. in the case of ion impact on polymers[1] it has been shown that large ionic fragments characteristic of the polymer are ejected. Thus it appears entirely feasible that the ion-impact technique coupled to a mass spectrometer should provide insight into the mechanisms of adsorption processes, particularly as one of the prime goals in catalytic chemistry is still the detection and quantitative measurement of the chemisorbed intermediates in catalytic reactions.

The examination of a prepared surface was first made by Honig.[2] He placed an ethylated germanium surface within a modified ion source of a mass spectrometer and observed the ejection of ions and neutrals characteristic of the surface when it was bombarded with 300 eV positive ions. Recently Fogel *et al.*, using an improved technique, have studied the catalytic decomposition of ammonia on platinum[3] and the synthesis of ammonia on iron.[4]

In this communication we report our preliminary findings on the synthesis of deutero-ammonia on a pure iron catalyst. This system was chosen for this initial study for two reasons. First, it allowed us to compare our results with those of Fogel *et al.*, whose technique differed to some extent in that they used higher ion impact energies (of the order of 20 keV), 60° sector mass spectrometry, continuous bombardment, continuous mass analysis, and a gas pressure of 10^{-4} torr. Secondly, despite the fact that the synthesis has been studied exhaustively by various other techniques over a number of years, there still remains sufficient controversy over the mechanism to warrant further study.

The conclusion reached by Fogel *et al.* was that the initial step is the adsorption of nitrogen in molecular form, followed by hydrogenation via a Rideal mechanism[5] to Fe-NH:

$$Fe-N_2 + H_2 \rightarrow 2Fe-NH$$

The third step is a direct hydrogenation to ammonia,

$$Fe-NH + H_2 \rightarrow Fe-NH_3$$

implying the absence on the surface of both adsorbed nitrogen atoms and amino groups. Our contrary views and those of other authors will become evident in our discussion.

Apparatus and technique

The basic apparatus used in this work has been previously described[6] and subsequently modified[7,8] to study both ion-molecule reactions in the gas phase and ion bombardment or sputtering of metals, salts, and polymers. Figure 1 shows schematically

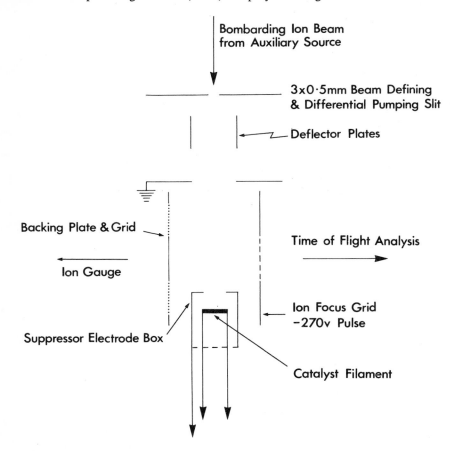

Fig. 1: Schematic diagram of the Bendix source, showing the primary beam direction and the catalyst region.

the principal features of the ion source and catalytic surface region. The iron catalyst was in the form of a filament (of cross area 2 x 0·15 mm, purity 99·998%, and supplied by Johnson Matthey, London) mounted via tungsten rods onto a movable non-magnetic stainless steel shaft which allowed (1) the surface to be adjusted for optimum secondary ion focusing at a few mm below the Bendix time-of-flight axis, (2) the surface to be positioned symmetrically between the backing plate and ion focus grid and (3) for ready withdrawal from the system, for filament replacement, via a vacuum lock. The filament was heated to a maximum of 900°C by d.c. current from a potential divider across a 12 V wet cell and the primary ion current (I_p) bombarding the catalyst surface was monitored on the analogue output system scanner (range $10^{-13}-10^{-6}$ A). I_p may be measured most accurately by means of a separate Faraday cage and suppressor electrode unit fitted onto a movable probe as described in reference 7. Effective monitoring of I_p may be made, however, by collection on the catalyst filament itself with suppression of secondary

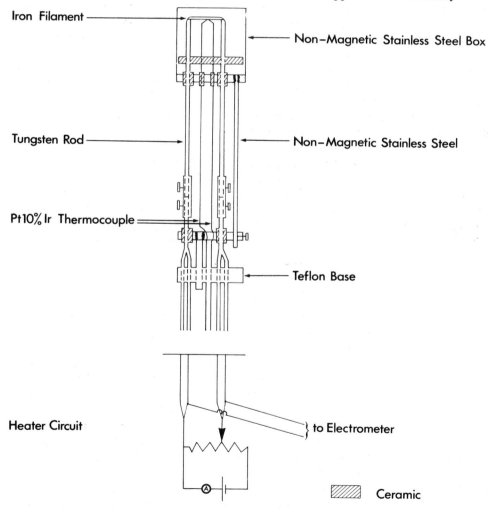

Fig. 2: Detail of the catalyst filament, thermocouple, and suppressor electrode.

electrons and negative ions, by application of a 40 V negative bias to the surrounding suppressor electrode containing the ion entry and exit slit of 1 x 5 mm. A typical bombarding ion intensity of 5 µA cm^{-2} was used at an impact energy of 2 keV.

The primary ion beam of singly-charged positive argon ions enters the ion source region from an external source and is focussed onto the catalyst surface by the deflector and focus plates shown in Fig. 1. Bombarding energies up to 2000 eV may be chosen. Secondary positive ions ejected from the catalyst surface are periodically sampled for analysis down the flight tube by application of the ion focus pulse.

The characteristic breakdown pattern due to charge transfer from the primary ion beam to the reacting gases may readily be obtained either by withdrawal of the catalyst surface from the source or by deflection of the primary to pass entirely in the gas phase on either side of the suppressor box (Figs. 1 and 2).

The technique used in the experiments to be described, depends on the physical measurement of the build-up of reactive species on a catalyst surface when a positive-ion inert gas beam is cut off for a given period of time. The build-up is observed as a rise above the steady state level previously established with a continuous bombarding ion beam.

The formation of deutero-ammonia from a 1:3 stoichiometric mixture of nitrogen and deuterium at a total pressure of $1 \cdot 1 \times 10^{-5}$ torr was studied. The major peaks monitored successively were those corresponding to N^+, ND^+, ND_2^+ and ND_3^+ at various catalyst temperatures in the range 55 to 907°C.

A brief examination of the effect of ion impact on ammonia adsorbed on the iron filament was made at temperatures of 100, 260 and 550°C by monitoring the same ion masses.

In experiments designed to investigate gas-surface interaction it is important to specify the condition of the surface being considered. A necessary requirement is that the surface is cleaned in a reproducible way. In this set of experiments a combination of three methods of surface cleaning were used. These in fact only keep surface contamination at a reproducible minimum as the background pressure of *ca.* 3-6 x 10^{-7} torr is rather higher than that required for an atomically clean surface. These methods are:

1   Thermal desorption
2   Surface reduction
3   Ion bombardment

In thermal desorption, the metal surface is heated in a vacuum to such a temperature that physically and chemically adsorbed molecules are desorbed. Wheeler[9] has found that the minimum temperature for complete thermal desorption, in °K, is given by $T_h \sim 20\Delta H$ where ΔH is the binding energy in kcal/mole of the layer of impurity atoms on the metal surface. Thus for the iron matrix, nitrogen is completely desorbed at 527°C and hydrogen at 367°C. Therefore, at a temperature of *ca.* 550°C, an iron surface should be clean at least with respect to nitrogen, hydrogen, water and some hydrocarbons. Thermal desorption of ferric oxide is impossible as it melts at 1,565°C which is above the melting point of iron, 1,535°C. However, any oxide present can be reduced with hydrogen as this reaction is thermodynamically favourable at the temperatures being considered here, particularly if a large excess of hydrogen is used.

Ion bombardment is also an effective surface cleaner, but as Yonts and Harrison[10] have stated, 'for a clean surface it is imperative that the beam particle current must be very much greater than the gas particle current'. Further, the degree of surface cleaning will also be dependent on the ion energy and hence efficiency of bombardment, i.e. one ion may remove up to ten particles.

Some experiments were carried out on surface cleaning and when a positive ion krypton beam of intensity $7\,\mu A/cm^2$ was used to bombard the iron surface it was found that in the absence of any reacting gases there was a maximum at *ca*. 340°C in the profile of secondary ion intensity v. temperature for most peaks. This was particularly the case for peaks $18\,(H_2O^+)$, $27\,(C_2H_3^+)$, $29\,(C_2H_5^+)$. As a result of this data, the following procedure was adopted for surface cleaning. The iron surface was first heated to a temperature of *ca*. 320°C for an hour in an atmosphere of *ca*. 10^{-5} torr of hydrogen. This reduces and/or desorbs any impurities on the surface. The hydrogen was then pumped out and the heated surface bombarded with a 2,000 eV argon positive ion beam of intensity *ca*. 5 $\mu A/cm^2$ for an hour. The surface was then ready for a series of runs.

Once the surface has been cleaned, the potential divider (see Fig. 2) is adjusted to give the temperature required for a run and the gas mixture allowed into the reaction chamber to a pressure of $1 \cdot 1 \times 10^{-5}$ torr. Approximately one hour is then allowed for the system to come to equilibrium before the delay experiments are carried out.

A partial spectrum is first run with a pen-recorder chart speed of $1 \cdot 7$ cm/min to obtain an idea of the background. The mass-selector on the secondary-ion electrometer is then set precisely on a particular mass with the selector switch on manual and the chart speed increased to 17 cm/min. This has the advantage of improving the sensitivity of the recorder trace.

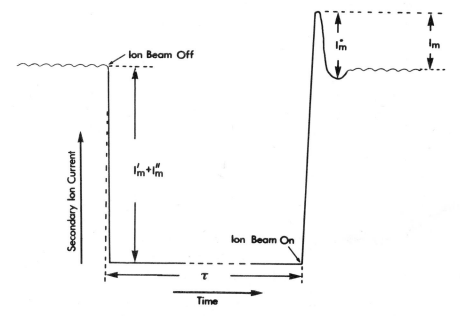

Fig. 3: Typical recorder trace of secondary ion current as a function of time.

After an equilibrium trace has been established for a particular mass-to-charge ratio, the primary ion beam is effectively cut off by deflecting the beam so that it does not pass the beam-defining slit (Fig. 1). On removal of the deflecting field, after a chosen time interval (τ), the secondary ion signal is found to rise above the equilibrium level and then return to it via a minimum. A schematic of the recorder trace is shown in Fig. 3. I_m'' is the contribution to the particular mass peak 'm' due to charge transfer from the primary ion beam to any of the reacting and background gases in the vicinity of the catalyst surface. I_m' is the contribution to the same mass peak due to ion-impact on the surface, and I_m is the increase in secondary ion signal observed after non-bombardment of the surface for the chosen time interval τ. This interval (generally 60 seconds) was of sufficient duration that I_m was independent of τ and a function only of reaction conditions and the current I_p. The increase I_m has been found to be quite reproducible, and the ratio of the minimum reading I_m^* to average reading I_m has been found to be constant for all mass peaks at all temperatures. I_m is taken to be proportional to the build-up of adsorbed species rapidly attained on the catalyst surface. The possibilities that I_m was (1) a function of background gases or (2) an electronic effect, were eliminated by repetition of the experiments in the absence of deuterium and nitrogen when in fact I_m approached zero.

Results

The $I_m(T)$ plots for each of the four ions N^+, ND^+, ND_2^+ and ND_3^+ are shown in Figs. 4 to 7 and it is apparent that, in the temperature range 55 to 900°C, there is a marked simil-

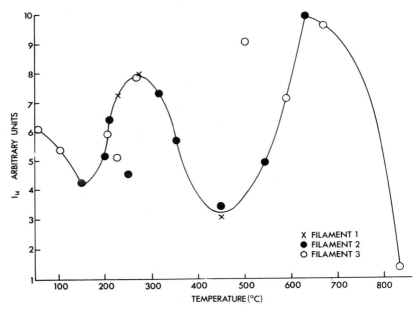

Fig 4: Dependence of I_m on temperature for the N^+ ion.

arity in the dependence of secondary-ion currents on temperature for each of the ions considered. Maximum yields are reached at 300 and 650°C (and possibly at 200°C) and minimum yields at 150 and 500°C (and possibly 250°C), these values being subject to error of ± 50°C. The comparative curves obtained by Fogel et al.,[4] for an active iron catalyst and covering the temperature range 0 to 800°C, show maxima for N^+, NH_2^+ and NH_3^+ at 400°C and minima for the same ions at 200°C. The major point of difference regarding these three ions is therefore our observation of a second maximum in the region of

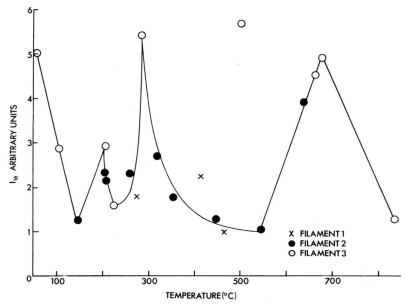

Fig. 5: Dependence of I_m on temperature for the ND^+ ion.

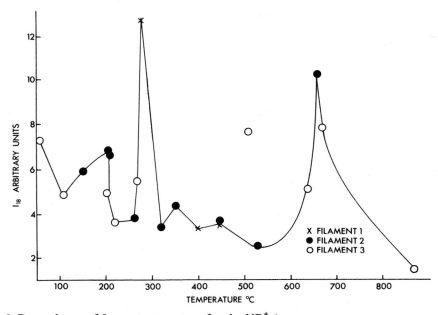

Fig. 6: Dependence of I_m on temperature for the ND_2^+ ion.

650°C whereas Fogel's yields decrease continuously from 400°C upwards. For the NH^+ ion, Fogel's and our data differ markedly. In Fogel's case the ion yield rose continuously from zero at about 100°C to a maximum at 300°C, descended to a minimum at 400°C, and

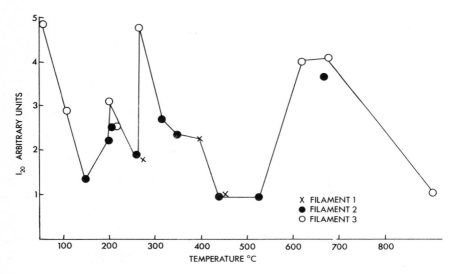

Fig. 7: Dependence of I_m on temperature for the ND_3^+ ion.

then rose almost linearly to 800°C (the maximum recorded). Figure 5 shows that our data has some similarity in the temperature range 150 to 650°C but is quite different above and below this range.

Correlation between the various $I_m(T)$ profiles was tested by plotting the ratios I_{14}/I_{20}, I_{16}/I_{20} and I_{18}/I_{20} against temperature. The least squares line and 95% confidence limits are shown in Figs. 8 to 10.

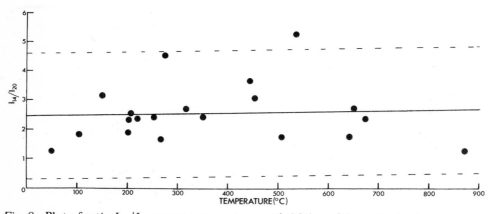

Fig. 8: Plot of ratio I_{14}/I_{20} versus temperature, with 95% confidence limits shown.

The temperature dependence of the three ratios is small as seen from the linear realtions (T being in °C):

$$I_{14}/I_{20} = 2 \cdot 44 + 2 \cdot 7 \times 10^{-4} \, T$$

$$I_{16}/I_{20} = 0 \cdot 89 + 3 \cdot 95 \times 10^{-4} \, T$$

$$I_{18}/I_{20} = 2 \cdot 30 - 3 \cdot 5 \times 10^{-4} \, T$$

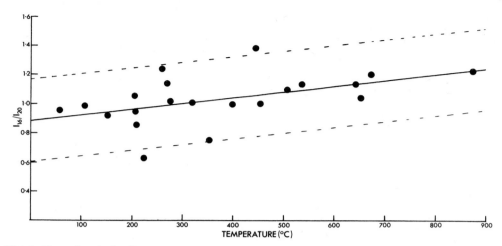

Fig. 9: Plot of ratio I_{16}/I_{20} versus temperature, with 95% confidence limits shown.

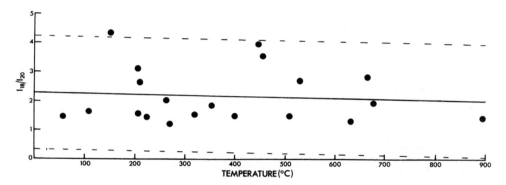

Fig. 10: Plot of ratio I_{18}/I_{20} versus temperature, with 95% confidence limits shown.

Thus at the typical synthesis temperature of 400°C the ratio of ion peaks

$$I_{14} : I_{16} : I_{18} : I_{20} \text{ is } 2.55 : 1.05 : 2.15 : 1.00$$

No detectable I_2 signal from adsorbed deuterium atoms was observed, and those from D_2 and N_2 proved too difficult to measure since their magnitude was less than 0.1% of the respective total $(I_m'' + I_m' + I_m)$ signals. An improved technique involving backing-off potentials should eliminate this difficulty in the future. The magnitude of the I_{Fe^+} and I_{Ar^+} signals were respectively 25 and 600 times the $I_{ND_3^+}$ signal approximately. We detected no ND_4^+, FeN^+, $FeND^+$, FeN_2^+, $ArND^+$, or $N_2D_n^+$ ($1 \leq n \leq 4$). FeN_2^+ was observed by Fogel *et al.* who suggested that the first step in the synthesis was $Fe + N_2 \rightarrow FeN_2$. The ion $ArNH^+$ has been observed in argon/ammonia ionised mixtures.[11] Melton and Emmett[12] on flash heating iron in ammonia observed the species N and NH and speculated that the compounds responsible were possibly Fe_4N, Fe_4NH and $FeNH$.

The independence of $I_m(T)$ from τ proves that equilibrium is rapidly attained on the surface. The further independence from temperature of several of the ion ratios corresponding to various intermediates suggests that the $I_m(T)$ profiles obtained are solely dependent on surface properties unrelated to the kinetic mechanism. For instance the high temperature maximum may possibly be correlated with the temperature variation of the specific

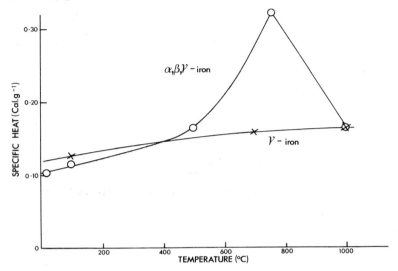

Fig. 11: Dependence of specific heat on temperature for γ-iron, and α-, β-, and γ-iron in equilibrium.

heat of iron. Iron consisting of α, β and γ forms in equilibrium has a maximum specific heat at 760°C (Fig. 11). Tovbin and Chalenko[13] have found that iron used to catalyse the ammonia synthesis undergoes a decrease in resistance. This they explained by suggesting that the synthesis of ammonia involves the formation of the γ-form of iron which is normally unstable at 400 to 500°C, the temperatures usually involved. Moreover the γ-iron gives a specific heat versus temperature curve which flattens markedly towards higher temperatures (see Fig. 11). Thus the increase in the proportion of γ-iron during the ammonia synthesis will move the observed maximum at 760°C to lower temperatures and this may well coincide with the observed maximum at 650°C.

A direct proportionality between the secondary ion current I_m and the build-up of surface concentration $[m_a]$ of the chemisorbed species is assumed according to the relation:

$$I_m = f \cdot \eta_m \cdot \gamma_i \cdot I_p \cdot [m_a]$$

where f is the fraction of the surface effectively cleaned of adsorbed species by the primary beam. This relation can be expected to apply only if the condition that the primary ion arrival rate and hence the sputtering rate is very much less than the rate of formation of any of the types of chemisorbed species. The term η_m is a combination of the collection efficiency of the ejected ion of type m and the multiplier detection efficiency for that ion. For the present we assume η_m to be equal for all ions although this is known not to be strictly true (see p. 207).

The probability γ_i that a chemisorbed species, of nominal mass m, will be sputtered on ion impact as an ion of mass m is here, for lack of contrary evidence, assumed to be equal for all types of mass and species. It should be borne in mind however that for metals, alloys, and polymers it has been established[1, 14] that the γ_i values are functions of the materials and do vary with bombarding ion type and impact energy.

Further the γ_i values at a particular argon ion energy are assumed to be independent of the surface temperature. This is reasonable on energetic grounds since for kilovolt sputtering the mean energies of the particles sputtered are of the order of 3 to 5 eV and a surface temperature of 800°C (equivalent to 0·1 eV) is unlikely to have a gross effect on the relative degree of ionisation or sputtering yields of the different adsorbed species.

A further necessary condition is that ion impact does not result in a significant degree of dissociation of the chemisorbed species. Dissociative charge transfer in the gas phase is a familar phenomenon. Argon ion impact on ammonia has been studied[11, 15, 16] and some collected data is presented in Table 1. The production of NH_2^+ is almost resonant, as is the first excited state of NH_3^+ at 15·31 eV. One might expect on sputtering (with 2 keV argon ions) ammonia molecules adsorbed on iron, that NH_2^+ would be a major product ion. By experiment however, using our present technique and an ammonia pressure of 2×10^{-6} torr we detected at three different temperatures (100, 260 and 550°C) a large yield of NH_3^+ ions but no NH_2^+. Thus it would appear that the chemisorbed species are ejected mainly by relatively low energy collisions within the surface.

Table 1. Dissociative charge transfer between argon ions and ammonia; relative abundances of the product ions as a function of ion impact energy.

	Ar^+ ion energy in eV										
	3^a	4^b	10^a	19^b	20^a	100^a	200^a	400^c	900^b	2000^d	60^e
NH_3^+	67	55·1	56	48·1	49	39	36	35	36·1	38·5	59
NH_2^+	33	44·9	43	51·7	50	60	63	63	63·6	60·8	37
NH^+	–	–	1	0·2	1	1	1	2	0·3	0·6	2
N^+	–	–	–	–				1		0·1	2

^a taken from curves in reference 11
^b reference 15
^c reference 16
^d this work
^e case of electron impact at 60 eV (from API tables)

The recombination energies of the majority of the argon ions in the bombarding beam are 15·76 eV ($^2P_{3/2}$ state) and 15·94 eV ($^2P_{1/2}$) and these are present in an abundance ratio of 2:1 since 40 eV electron energy was employed in the ion source. The ionisation and appearance potentials of various particles of interest are listed in Table 2. The ionisation potential values for ND_2 and ND are estimated as the ionisation potentials of NH_2 and NH respectively plus the difference in the ionisation potentials of ND_3 and NH_3. On the basis of the near-adiabatic theory,[21] the reaction cross section between argon ions (at the relatively low translational energies considered here) and the four ND_n ($0 \leq n \leq 3$) species in the gaseous state could be expected to be in the order $\sigma_{ND} > \sigma_N > \sigma_{ND_2} > \sigma_{ND_3}$. On the basis of direct interaction between argon ions and these species adsorbed on the

Table 2. Ionisation and appearance potentials of nitrogen, hydrogen, ammonia and their fragments.

Species	Ionisation potential (eV)	Ref.	Ion		Appearance potential (eV)	Ref.
NH_3	10.15	18	NH_2^+	(from NH_3)	15.8	31
ND_3	11.52	17	NH^+	(from NH_3)	19.5	31
NH_2	11.2	18				
ND_2	12.5	*	N^+	(from NH_3)	25.0	31
NH	13.9	19				
ND	15.2	*				
N	14.54	20	N^+	(from N_2)	24.3	31
N_2	15.55	31				
H_2	15.43	31	H^+	(from H_2)	18.05	31
H	13.6	31				

* estimated

surface we might expect a similar dependence. However this view ignores the effect of (1) the surface bonding and (2) the probability that the impinging particle is not an argon ion but a secondary charged or uncharged particle of argon, iron, or a nitrogen-containing species involved in the energy cascade. It has been established that for impact energies of a few keV, the mean energies of both atomic[14] and molecular ions[1] sputtered is below 10 eV and generally in the range 2 to 5 eV. The value of the ionisation potential of the species does play a major role in the case of secondary ionisation from alloys. Dillon et al.[22] have shown that the dependence is approximately $I_m \propto e^{I/kT}$ where I^* is a value close to the first ionisation potential of the metallic constituent of the alloy. Thus if one applies this result to the chemisorbed species the expected order of sputtering yields would be $\gamma_{ND_{3a}^+} > \gamma_{ND_{2a}^+} > \gamma_{N_a^+} > \gamma_{ND_a^+}$. Thus the two effects tend to annul each other. There is however no quantitative evidence on the ion yields of adsorbed species.

The fact that the fraction $I_m/(I_m + I_m' + I_m'')$ for the case of N^+ ions is as high as 0.2 strongly suggests that nitrogen atoms are present as adsorbed species on the surface. However a possible source of mass 14 is the process $N_{2a} \rightarrow N^+$ on sputtering with 2 keV argon ions. In the gas phase such dissociative charge transfer is very inefficient, the proportion of N_2^+ being 99.9% and N^+ 0.1%. The principle reason is the large energy deficit of ca. 10 eV between the recombination energy of the argon ion and the appearance potential of N^+ from N_2. However it has been previously noted that, in a more favourable case, there is no dissociative ionisation of adsorbed ammonia on argon ion impact at 2 keV.

To test whether nitrogen dissociated on an iron catalyst, Fogel et al. bombarded an active filament in a nitrogen pressure of ca. 10^{-4} torr and measured the ion yields for N^+ and N_2^+ over the temperature range 0 to 800°C. The two curves show a similar form and Fogel et al. concluded that no dissociation occurred. On replotting their data in the form of $I(N^+)/I(N_2^+)$ against temperature it becomes apparent that the ratio decreases with temerature rise to a minimum in the region of 300°C and then rises by more than 100% in the interval 300 to 800°C. Such behaviour does not, we think, preclude nitrogen dissociation.

The finding of Kummer and Emmett[23] that the rate of the synthesis of ammonia was comparable to the rate of isotopic exchange for $^{28}N_2 + ^{30}N_2 \rightleftharpoons ^{29}N_2$ suggests that adsorbed nitrogen atoms are present. Tamaru[24] has however drawn attention to the mutual influence of hydrogen and nitrogen on their rates of chemisorption which possibly invalidates the above comparison. Scholten et al.[25] concluded that the rate determining step is the chemisorption of nitrogen and hence that all further hydrogenation steps may be considered at equilibrium. On the basis of measurements of the rates of adsorption and desorption as a function of both temperature and coverage, the chemisorption isotherm and differential entropies were calculated. Their interpretation is that at low coverages an atomic immobile adsorption occurs but that, at higher coverages in the range important for normal synthesis conditions, the nitrogen is adsorbed in the form of partially dissociated molecules (–N = N– or = N–N=) which move freely over the surface.

Our conclusion is therefore, assuming equal γ_i values for the present, that the surface concentrations of the four nitrogen containing species are in the ratio

$$N : ND : ND_2 : ND_3 \text{ as } 2.55 : 1.05 : 2.15 : 1.00$$

at a temperature of 400°C.

Thus we oppose the suggestion of Ozaki et al.[26] that the imine (NH) radical is the major species on the surface and hence that the equilibrium determining the amount of nitrogen on the surface would be

$$NH_a + H_2 \rightleftharpoons NH_{3a}$$

Logan and Philp[27] had suggested on a reanalysis of the data of Ozaki et al. that their conclusion was erroneous and that their evidence is more consistent with the major presence of N atoms on the surface combined with the action of a H/D kinetic isotope effect in the dissociative adsorption of nitrogen.

Mars et al.[28], on the basis of some approximate energy calculations, have suggested that the monohydrogenated surface complex Fe-NH-NH-Fe has a low energy and hence, assuming the system is at equilibrium, will occur frequently. However neither Fogel et al. nor the present authors have observed the $N_2H_2^+$ ion or any other hydrogenated ion containing two nitrogen atoms. That two or more bonds may be broken in a molecular complex as a result of a single ion impact is known from sputtering studies on polymers[1] where a first order dependence of the yields of large molecular fragments on ion bombarding current occurs.

The adsorbed nitrogen atoms can be reduced to the imine group (ND) in either of two ways:

(a) via a Rideal mechanism[5] in which the adsorbed nitrogen atom is reduced by molecular deuterium loosely bound to the surface or arriving from the gas phase

$$2N_a + D_2 \rightleftharpoons 2ND_a$$

or
(b) via a Langmuir-Hinshelwood mechanism[32] in which reaction occurs between two species adsorbed as atoms on the surface:

$$N_a + D_a \rightleftharpoons ND_a$$

Thus this mechanism requires that deuterium is dissociatively adsorbed on the iron surface:

$$D_2 \rightleftharpoons 2D_a$$

The rapid isotopic exchange of hydrogen-deuterium mixtures on iron catalysts at temperatures far below the normal synthesis temperature strongly suggests that hydrogen is dissociatively adsorbed. Fogel et al. however, on the basis of ion impact studies on hydrogen adsorbed on iron, state that no dissociation occurs. Examination of their data shows that the ratio

$$\frac{I(H^+)}{I(H_2^+)}$$

does not stay constant, or even rise slightly with temperature as might be expected on energetic gounds, but decreases markedly with rise of temperature. This result may be due to a combination of the effects of dissociation, desorption and diffusion. ND_a may be converted to ND_{3_a} in possibly four ways:

(a) The reduction of the ND on the surface directly to ND_3 by loosely bound molecular deuterium using a Rideal mechanism

$$ND_a + D_2 \rightleftharpoons ND_{3_a}$$

(b) A two stage reduction by adsorbed deuterium atoms using a Langmuir-Hinshelwood mechanism

$$ND_a + D_a \rightleftharpoons ND_{2_a}$$
$$ND_{2_a} + D_a \rightleftharpoons ND_{3_a}$$

(c) A single stage reduction by adsorbed deuterium atoms in a three-body molecular reaction

$$ND_a + 2D_a \rightleftharpoons ND_{3_a}$$

(d) A single stage reduction by adsorbed deuterium atoms to yield $Fe-ND_2$, followed by a disproportionation reaction of the adsorbed amino group.

$$ND_a + D_a \rightleftharpoons ND_{2_a}$$
$$2ND_{2_a} \rightleftharpoons ND_{3_a} + ND_a$$

Reaction (c) is not favoured on energetic grounds, but it is principally rejected, together with reaction (a) on the evidence that the amino group is present in the adsorbed state on the surface at a comparable concentration to ND and ND_3. Thus two possible reaction schemes remain and the one based on the second alternative is:

$$D_2 \rightleftharpoons 2D_a \qquad (0)$$

$$\tfrac{1}{2}N_2 \underset{k_{-1}}{\overset{k_1}{\rightleftharpoons}} N_a \qquad (1)$$

$$N_a + D_a \underset{k_{-2}}{\overset{k_2}{\rightleftharpoons}} ND_a \qquad (2)$$

$$ND_a + D_a \underset{k_{-3}}{\overset{k_3}{\rightleftharpoons}} ND_{2a} \qquad (3)$$

$$ND_{2a} + D_a \underset{k_{-4}}{\overset{k_4}{\rightleftharpoons}} ND_{3a} \qquad (4)$$

$$ND_{3a} \rightleftharpoons ND_3 \qquad (5)$$

In this scheme a closed sequence is being considered, in which the active centres are regenerated. Thus, according to the steady state approximation, each of the overall reaction rates for the elementary reactions will be equal

$$r_1 = r_2 = r_3 = r_4 = r_5 = r$$

and hence we have the relations

$$k_1 [N_2]^{1/2} - k_{-1}[N_a] = r \qquad (i)$$

$$k_2 [N_a][D_a] - k_{-2}[ND_a] = r \qquad (ii)$$

$$k_3 [ND_a][D_a] - k_{-3}[ND_{2a}] = r \qquad (iii)$$

$$k_4 [ND_{2a}][D_a] - k_{-4}[ND_{3a}] = r \qquad (iv)$$

Substituting the observed I_m values for N^+, ND^+ ND_2^+ and ND_3^+ at 400°C into Eqns. (ii) to (iv) we obtain

$$2\cdot55\, k_2 [D_a] - 1\cdot05\, k_{-2} = r' \qquad (v)$$

$$1\cdot05\, k_3 [D_a] - 2\cdot15\, k_{-3} = r' \qquad (vi)$$

$$2\cdot15\, k_4 [D_a] - 1\cdot00\, k_{-4} = r' \qquad (vii)$$

where r'/r equals a proportionality constant relating secondary ion signals to the absolute surface concentrations.

Hence at 400°C, $[D_a] = \dfrac{1\cdot05\, k_{-2} - 2\cdot15\, k_{-3}}{2\cdot55\, k_2 - 1\cdot05\, k_3} = \dfrac{1\cdot00\, k_{-4} - 2\cdot15\, k_{-3}}{2\cdot15\, k_4 - 1\cdot05\, k_3}$ (viii)

By inspection, $k_{-2} \simeq k_{-4}$ and $k_2 \simeq k_4$

This similarity is possibly a reflection of the fact that in reactions (2) and (4) a deuterium atom is being added in each case to essentially stable species, namely N_a and ND_{2_a}. Reaction (3) appears twice as fast as reaction (2) or (4) and half as fast in the reverse direction. This might be expected as with the formation of ND_{2_a} a potential well is reached according to the approximate energy calculations of Mars et al.[28]

The alternative reaction sequence is one where reactions (0) to (3) and (5) remain the same and reaction (4) is replaced by reaction (6)

$$2ND_{2_a} \rightleftharpoons ND_{3_a} + ND_a \qquad (6)$$

According to the steady state condition we have

$$\dfrac{d}{dt}[ND_a] = \dfrac{d}{dt}[ND_{2_a}] = \dfrac{d}{dt}[ND_{3_a}] = 0 \qquad (ix)$$

hence $k_2 [N_a][D_a] + k_{-3}[ND_{2_a}] + k_6 [ND_{2_a}]^2 =$

$$k_{-2}[ND_a] + k_3[ND_a][D_a] + k_{-6}[ND_a][ND_{3_a}] \qquad (x)$$

and

$k_3[ND_a][D_a] + k_{-6}[ND_{3_a}][ND_a] = k_{-3}[ND_{2_a}] + k_6[ND_{2_a}]^2$ (xi)

By addition of (x) and (xi) we have

$$k_2[N_a][D_a] = k_{-2}[ND_a] \qquad (xii)$$

Substituting the relative values of $[N_a]$ and $[ND_a]$ from the secondary-ion data, we obtain

$$2\cdot55\, k_2 [D_a] = 1\cdot05\, k_{-2}$$

$$[D_a] = 0\cdot41\, k_{-2}/k_2 \qquad (xiii)$$

Thus the surface concentration is dependent only on reaction (2). There is at present no way of distinguishing between the two schemes although the first one where there is stepwise addition of deuterium atoms throughout the synthesis is the one generally accepted.[26-30]

Conclusion

The potentiality of the ion probe technique for investigating surface reactions has been demonstrated. Information on the relative concentrations of all the chemisorbed species that occur on a catalyst surface is feasible. The principle uncertainty is the relative ionisation probabilities for each type of species.

In the synthesis of deutero-ammonia on pure iron, the temperature variations of the adsorbed species, N, ND, ND_2 and ND_3, have been shown to be closely similar over the temperature range 55 to 900°C. Their relative values at 400°C, assuming equal ionisation probabilities, are in the ratio 2·55:1·05:2·15:1·00 respectively. The formation of the ion FeN_2^+ has not been confirmed and no ions characteristic of any hydrogenated species containing two nitrogen atoms have been detected. Two feasible kinetic schemes have been analysed, using the steady state approximation. Further progress will necessitate the determination of the adsorbed deuterium atom concentration and a more detailed examination of the decomposition of ammonia on pure iron.

References

1. A. F. Dillon, R. S. Lehrle, J. C. Robb and D. W. Thomas, *Advan. Mass Spectrom.* **4**, 477 (1968).
2. R. E. Honig, *Advan. Mass Spectrom.* **2**, 25 (1963).
3. Ja. M. Fogel, B. T. Nadikto, V. F. Ribalko, R. P. Slabospitskii, I. E. Korobchanskaja and V. I. Shvachko, *J. Catalysis* **4**, 153 (1965).
4. V. I. Shvachko, Ya. M. Fogel and V. Ya. Kolot, *Kinetica i Kataliz* **7**, 834 to 840 (1966).
5. D. D. Eley and E. K. Rideal, *Proc. Roy. Soc. (London) Ser. A.* **178**, 429 (1941).
6. R. S. Lehrle, J. C. Robb, D. W. Thomas, *J. Sci. Instr.* **39**, 458 (1962).
7. J. B. Homer, R. S. Lehrle, J. C. Robb and D. W. Thomas, *Advan. Mass Spectrom.* **2**, 415 (1964).
8. D. W. Thomas, 'Reaction of ions with molecules in the gas phase and the sputtering of a solid surface by ion bombardment', Chap. 7 in *Time-of-Flight Mass Spectrometry* (Eds. D. Price and J. E. Williams), Pergamon Press, Oxford, 1969.
9. A. Wheeler in *Structure and Properties of Solid Surfaces* (Eds. R. Gomer and C. S. Smith), 1953, p. 439.
10. O. C. Yonts and D. E. Harrison, *J. Appl. Phys.* **31**, 1583 (1960).
11. W. S. Koski and G. R. Hertel, *J. Am. Chem. Soc.* **86**, 1683 (1964).
12. C. E. Melton and P. H. Emmett, *J. Phys. Chem.* **68**, 3318 (1964).
13. M. V. Tovbin and V. G. Chalenko, *Ukr. Khim. Zh.* **31**, (7), 663 (1965)
14. M. Kaminsky, *Atomic and Ionic Impact Phenomena on Metal Surfaces*, Springer Verlag, 1965.
15. G. Sahlstrom and I. Szabo, *Arkiv. Fysik* **38**, 145 (1967).
16. G. K. Lavrovskaya, M. I. Markin and V. Talroze, *Kinetika i Kataliz* **2**, 18 (1961).
17. H. Neuert, *Z. Naturforsch.* **7A**, 293 (1953).
18. V. H. Dibeler, J. A. Walker, N. H. Rosenstock, *J. Res. Nat. Bur. Std.* **70A**, 459 (1966).
19. W. C. Price, T. R. Passmore and D. M. Roessler, *Discussions Faraday Soc.* **25**, 201 (1963).
20. C. E. Moore, *Nat. Bur. Std.* Circ. No. 467 (1949).
21. Massey and Burhop *Electronic and Ionic Impact Phenomena*, Oxford, 1952, p. 441.
22. A. F. Dillon, J. C. Robb and D. W. Thomas, in press.
23. J. T. Kummer, P. H. Emmett, *J. Phys. Chem.* **56**, 258 (1952).
24. K. Tamaru, *Trans. Faraday Soc.* **59**, 979 (1963).
25. J. J. F. Scholten, P. Zwietering, J. A. Konvalinka, J. H. de Boer, *Trans. Faraday Soc.* **55**, 2166 (1959).
26. A. Ozaki, H. S. Taylor, M. Boudart, *Proc. Roy. Soc. (London) Ser. A.* **258**, 47 (1960).

27 S. R. Logan and J. Philp, *J. Catalysis* **11**, 1 (1968).
28 P. Mars, J. J. F. Scholten, P. Zwietering, in J. H. De Doer (Ed.) *Mechanism of Heterogeneous Catalysis,* 1960, p. 68.
29 M. Temkin and V. Pyzhev, *J. Phys. Chem. (U.S.S.R.)* **13**, 851 (1939).
30 J. Horiuti and N. Takezawa, *J. Res. Inst. Catalysis, Hokkaido Univ.* **8**, 170 (1961).
31 F. H. Field and J. L. Franklin, *Electron Impact Phenomena,* Academic Press, 1957.
32 I. Langmuir, *Trans. Faraday Soc.* **17**, 621 (1921).

Discussion

Dr. van Heek: Are the experiments performed in isothermal steps?

D. Terrell: Each individual reading is taken under isothermal conditions after the system had reached equilibrium.

Dr. van Heek: Do you use deuterium to avoid overlap with the water peaks?

D. Terrell: Deuterium was used to give a peak separation of two mass units for the species being monitored. Overlap with the water peak at $m/e = 18$ does not arise since this is incorporated in the background level. No intensity rise is observed for the background $m/e = 18$ peak.

Chapter 8

Studies of Negative Ion Formation at Low Electron Energies

P. W. Harland, K. A. G. MacNeil and J. C. J. Thynne

Chemistry Department, Edinburgh University, Scotland

Introduction

Electron bombardment of a molecule may result in the formation of positive and negative ions. The latter may be produced by three different processes:

(i)   resonance attachment
$$AB + e \rightarrow AB^-$$
(ii)   dissociative resonance capture
$$AB + e \rightarrow A^- + B$$
(iii)   ion-pair formation
$$AB + e \rightarrow A^- + B^+ + e$$

These mechanisms usually operate at different electron energies, the resonance processes usually occurring in the 0-10 eV energy region and the ion-pair processes at energies above this.

In electron impact studies, when the source of the electrons is a heated filament, uncertainties arise in the interpretation of the experimental ionisation efficiency curves because of the energy spread of the thermionically emitted electron beam. As a result, the ionisation thresholds become indeterminate or smeared-out because of the high energy tail of the electron energy distribution. In addition, any structure in the ionisation probability curve (which is generally narrow compared to the distribution width) is obscured.

Various methods have been used in an attempt to remove or reduce the effect of this quasi-Maxwellian electron energy distribution. An experimental solution has been to use the 'retarding-potential-difference' method.[1-3] This involves the removal of a narrow energy slice from the distribution by applying a retarding potential and measurement of the ion current before and after the removal. This technique, although used by several workers with much success, is not always simple; use of a thin slice so as to obtain essentially monoenergetic electrons often reduces the intensity of the electron current drastically with an adverse effect on the signal-to-noise ratio for the measured ion currents. This is especially a problem when negative ions are studied since these are frequently of very low intensity.

Another approach has been to use analytical methods to remove the energy spread. Morrison[4] has applied such a technique to the problem of positive ions. He showed, from tests made on several artificial examples and on two actual cases, that the method was promising and revealed structure in ionisation efficiency curves.

In any study of negative ions the energy spread problem is critical, particularly when dissociative resonance processes occurring at low energies are being examined. The energy spread causes uncertainty in the onset energy; perhaps more important, if several such ion-formation processes occur within a small energy range, broadening of the individual resonance peaks occurs resulting in overlap. Consequently, experimental observations may indicate only one wide resonance peak, thereby obscuring the individual processes responsible.

We have therefore attempted to obtain a solution to this problem for negative ions by applying the existing analytical methods to remove the effect of the electron energy distribution; we have assessed the reliability of this solution and applied our findings to results obtained from an experimental study of the negative ions formed by sulphur dioxide, carbonyl fluoride, hexafluoroacetone and pentafluorosulphur chloride.

Experimental

The mass spectra were obtained using a Bendix time-of-flight mass spectrometer, Model 3015. The electron energy was measured using a Solartron digital voltmeter LM 1619. Using two channels of the mass spectrometer analogue output scanners enabled two mass peaks (e.g. O^-/SO_2 and F^-/CF_2O) to be measured simultaneously on 1 mV Kent potentiometric recorders. This was of particular value in energy scale calibration procedures since no switching between peaks was required.

The electron current was kept constant automatically over the energy range studied; a very small electron current was used ($\sim 0.0025\,\mu A$) in order that the filament temperature was kept low, so as to minimise any effects due to thermal decomposition of the substances studied on the filament. Ion source pressures were usually maintained below 5×10^{-6} mm Hg.

The ionisation curves for the major ions were normally measured at least five times; the appearance potentials for the thresholds were (after deconvolution) reproducible to at least ± 0.1 eV, often better.

The appearance potential of the O^- ion from SO_2 was used as the reference for energy scale calibration,[3] both the onset at 4.2 eV and the maximum of the resonance peak at 5.0 eV being taken as the calibration points.[5] The electron energy distribution, which was required to be known for the deconvolution procedure, was measured using the SF_6^- ion formed by sulphur hexafluoride.[6,7] It was found that performing 15 smoothing and 20 unfolding iterations on the basic experimental data enabled satisfactory evaluation of appearance potentials, resonance peak maxima and peak widths (at half-height), to be made.

Materials. Sulphur hexafluoride (Air Products), sulphur dioxide (BDH) and carbonyl fluoride (Pierce Chemical Co.), were obtained from cylinders and used without further purification.

Hexafluoroacetone was obtained by dehydration of the sesquihydrate. The impurities were fluoroform, hexafluoroethane and carbon dioxide and these were removed by by prolonged pumping on a vacuum line at $-130°C$.

Pentafluorosulphur chloride was a gift from Dr. H. L. Roberts (ICI Mond Division).

Method of deconvolution

The techniques used for carrying out the deconvolution process have been discussed by Morrison[4] and others.[8-10] In the work reported here, we required to know to what extent appearance potentials, resonance peak maxima and the peak widths obtained using these methods could be considered reliable. In addition, the effects of random noise in the experimental data and of inexact estimates of the form of the electron energy distribution upon the reliability of the results was required to be known.

Accordingly, artificial examples have been constructed and deconvoluted under various noise level conditions and with a variety of electron energy distributions. For the sake of clarity a summary of our application of the Morrison-van Cittert method[4, 8] is given here.

Thermionically emitted electrons possess an energy distribution such that the probability of a particular electron having an energy in the range U to $U + \Delta U$ is given by $m(U)$, where:

$$\sum_{U=0}^{\infty} m(U) \Delta U = 1$$

Before impact, the electrons are accelerated by a potential difference V, and emerge from the electron gun with an energy E, where $E = U + V$. If the probability of the formation of a particular ion by an electron of energy E is a function of E, namely $I(E)$, then the probability of formation of the ion by a beam of electrons of nominal energy V is $i(V)$, where:

$$i(V) \propto \sum_{U=0}^{\infty} I(V+U) \, m(U) \, \Delta U \qquad (1)$$

The experimental procedure is to sample the ion current $i(V)$ at equal intervals of V and required to be found are the corresponding values of $I(E)$, which satisfy Eqn. (1).

When the electron energy distribution $m(U)$ is known at the same intervals in energy, and $i(V)$ and $m(U)$ are bounded by zero at the upper and lower ends of the energy range, it is possible to obtain unique solutions for $I(E)$ at comparable intervals in E.

One method of solution for this process is that of van Cittert,[8] where the set of iterative equations:

$$I(E)_{calc(n+1)} = I(E)_{calc(n)} + i(V) - \sum_{U=0}^{\infty} I(E)_{calc(n)} \, m(U) \Delta U \qquad (2)$$

tends to $I(E)$ as n increases. This method takes a trial function for $I(E)$, convolutes it with the electron energy distribution and compares the results with the original $i(V)$. The difference is then used as a correction to the trial functions and the procedure is repeated. The original trial function used is the set of $i(V)$ itself, i.e. when $n = 0$, $I(E)_{calc(n)} = i(V)$.

However, as Morrison[4] has pointed out, if this process is applied to actual experimental data, the solution obtained for $I(E)$ is almost so obscured by noise as to be effectively useless. Smoothing of the original data is necessary before the deconvolution operation may be performed. The following iterative process removes the random scatter very effectively:

$$i(V)_{calc(1)} = \sum_{U=0}^{\infty} i(V)_{obs} \, m(U) \, \Delta U$$

$$i(V)_{\text{calc }(n+1)} = i(V)_{\text{calc }(n)} + \sum_{U=0}^{\infty} [i(V)_{\text{obs}} - i(V)_{\text{calc }(n)}] m(U) \Delta U \quad (3)$$

In addition, if at any stage in these operations (2) and (3) $I(E)$ goes negative then it may be set equal to zero.

Application of the method to artificial systems

In order to examine the applicability of these methods to our experimental system, initially known functions were convoluted with a known electron energy distribution and

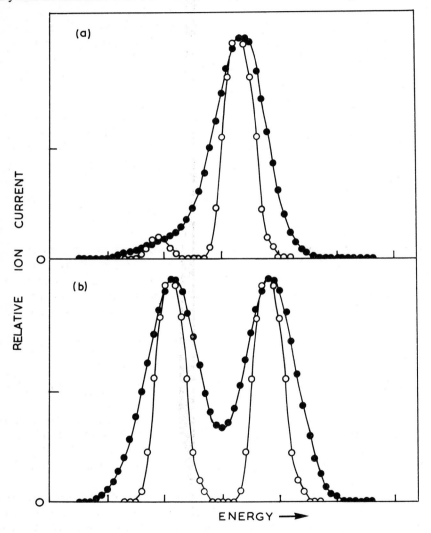

Fig.1: Artificial functions before and after convolution.
(a) shows function A, O before convolution, ● after convolution.
(b) shows function B, O before convolution, ● after convolution.
Convolution shifts the functions to lower energies but the peaks have been re-aligned for ease of presentation.

then deconvoluted. Equation (1) was used for the convolution process, the upper limit of the summation in practice being finite. Computer programmes were written to perform the smoothing and deconvolution processes.

In Figs. 1 a and b we show the two artificial functions (A) and (B) used before and after convolution. The electron energy distribution used was that shown in Fig. 7 a. These artificial functions were chosen because, when convoluted, they showed some resemblance to the experimental ionisation curves we have observed in these studies. The electron energy distribution was measured experimentally using the SF_6^- ion from SF_6, since it has been shown that the energy dependence of formation of this ion is very closely related, on a reversed energy scale, to the electron energy distribution.[6,7] It is apparent from Fig. 7 a

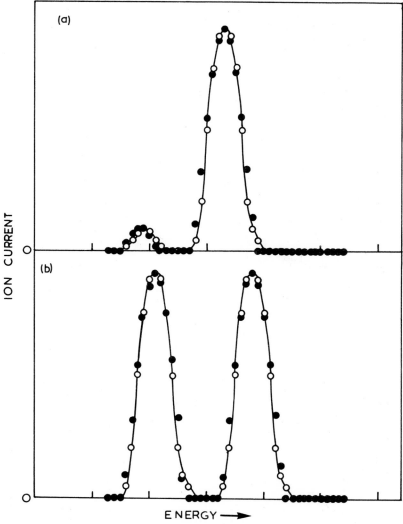

Fig. 2: Deconvolution of artificial function; results of 15 smoothing and 20 unfolding iterations on the convoluted forms of (a) function A (b) function B. Original data indicated by open circles; full circles indicate the unfolded results.

that the distribution is fairly symmetrical, about 44% of the electrons being on the low energy side. The width at the baseline is about 3 eV and at half-peak-height about 0.9 eV.

In Figs. 2 a and b the results of performing 15 smoothing iterations and 20 unfolding iterations upon functions (A) and (B) are shown, together with the original functions. It is apparent that the positions of the ionisation thresholds, the peak maxima and the peak widths together with the relative intensities are accurately recovered.

However, these results are obtained under experimentally unrealisable conditions; in practice experimental data contain random scatter in the individual points. As a test of a system corresponding more nearly to an experimental one, random noise was added to the convoluted functions shown in Fig. 1, and the deconvolution operation repeated. Random noise was added by generating a set of random numbers with a normal distribution, multi-

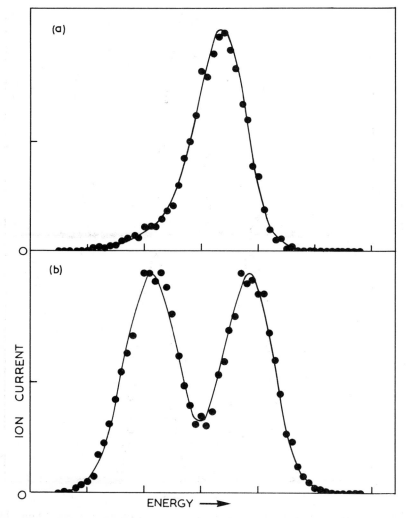

Fig. 3: Convoluted forms of functions A and B with random noise added (higher level). The unbroken line indicates the noiseless function.

plying these by the square root of the appropriate $i(V)$ and then by a constant and adding this value to $i(V)$. Two levels of noise were added, one being approximately twice the other. Figures 3 a and b show the synthetic convoluted function with the higher level of noise added: the full line in Fig. 3 refers to the smooth forms. It should be noted that this level of noise is somewhat greater than found in our experimental work.

The results of performing 15 smoothing and 20 deconvoluting iterations on these data are shown in Fig. 4 together with the original functions. The results for the lower noise level are shown in Fig. 4 a and c and for the higher noise level in Fig. 4 b and d. It is clear that for both artificial functions, at the lower noise level, the energy values for the

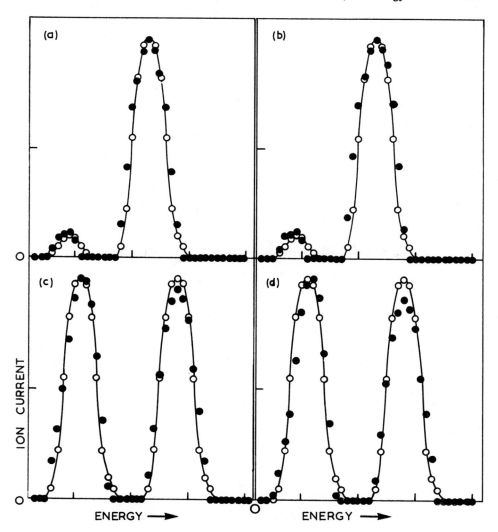

Fig. 4: Deconvolution of 'noisy' convoluted forms of artificial functions; 15 smoothing and 20 unfolding iterations performed. Function A results are shown in (a) low noise level and (b) higher noise level. Open circles : original data, full circles : unfolded results; (c) and (d) show similar results for function B.

onsets, peak maxima and the widths of the peak are again accurately recovered. In the case of function (B) it appears that there is a small change (∼5%) in the relative peak heights.

At the higher noise level, for function (A) the onsets and the maxima for both peaks are identical, but there is a slight broadening of the major peak and a small change in the relative intensities of the peaks. For function (B) the small change noted in the relative peak heights at the low noise level has become more apparent at this higher level, and there is a shift of one energy interval in the positions of the first onset and the peak maximum. This shift corresponds to 0·1 eV experimentally.

In general, it is seen that the effects of random scatter of the data upon the deconvolution operation can be dealt with effectively if a smoothing procedure is used. The considerable loss of resolution and information that results when smoothing is not used to

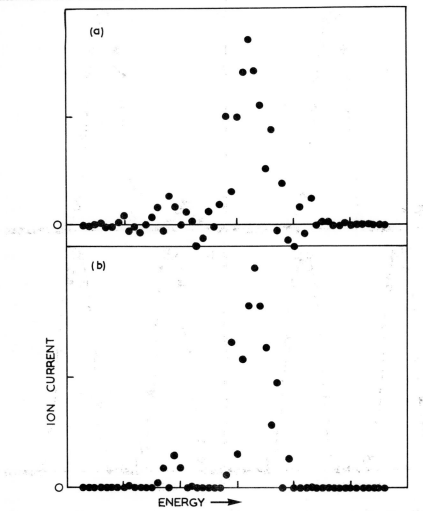

Fig. 5: Deconvolution of function A with lower level of random noise, 20 unfolding iterations only. In (a) no smoothing; (b) negative values of $I(E)$ have been set to zero whenever they occurred in the deconvolution process.

remove noise can be seen in Fig. 5. In Fig. 5 a is shown the results of 20 unfolding iterations upon the data for function (A) with the lower noise level added. The shape of the original function, although qualitatively apparent, is lost as regards yielding information concerning thresholds, peak widths etc.

A considerable improvement in the results is noted (see Fig. 5 b) when negative values of $I(E)$ are set equal to zero during deconvolution, but comparison of Figs. 5 b and 4 a indicate the necessity for using smoothing operations if accurate ionisation data are required.

In the results reported above we have used 20 deconvoluting iterations in each case. The effect upon the shape of function (A) of using different numbers of unfolding can be seen in Fig. 6 a-d. After only 4 deconvolutions the two peaks are distinguished and

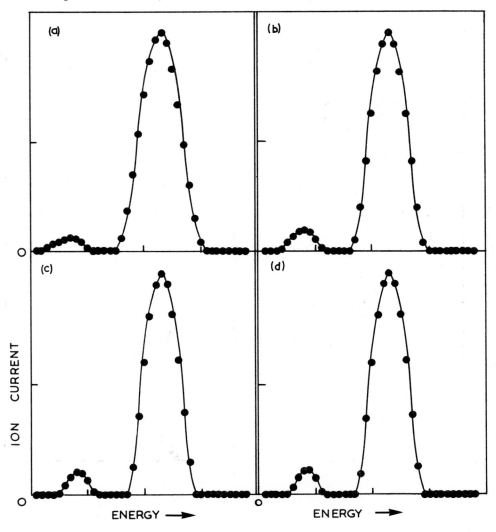

Fig. 6: Effect of variation of the number of unfolding iterations on the resolution of function A. (a) 4 (b) 10 (c) 20 and (d) 30 unfoldings.

resolved; both peaks are broader than in the original function and their thresholds are displaced to lower electron energy. Increasing the number of unfoldings decreases the resonance peaks widths, correctly establishes the relative peak heights; also sharper and more accurate onset energies are obtained. A noticeable improvement occurs by increasing the number of unfoldings from 10 to 20, but from 20 to 30 the difference is only marginal. Accordingly, in the work reported here we have normally used 16-20 deconvolutions to analyse our data. Similarly we have found that 15 smoothing iterations gives satisfactory convergence.

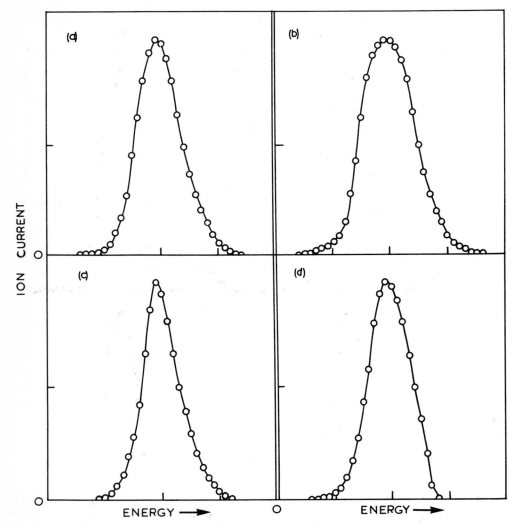

Fig. 7: The electron energy distribution used in the calculations.
    (a) Correct form of distribution (based upon SF_6^-),
    (b) Overestimate,
    (c) Underestimate,
    (d) Correct form of distribution with a severely curtailed high energy portion.

Morrison[4] has suggested that, for positive ions, quite large variations in the electron energy spread $m(U)$ have little effect upon the resolution of structure in an ionisation efficiency curve when the results are deconvoluted. We have examined the effect of poor estimates of the electron energy distribution upon our data. In our work this distribution is obtained experimentally using the SF_6^- ion formation from SF_6 since this has been shown to reflect accurately the energy distribution;[6,7] this 'correct' distribution is shown in Fig. 7a. Distribution 2 (curve b) is an overestimate, distribution 3 (curve c) an underestimate and distribution 4 (curve d) the 'correct' form as based upon SF_6^-, but with a severely curtailed high energy portion.

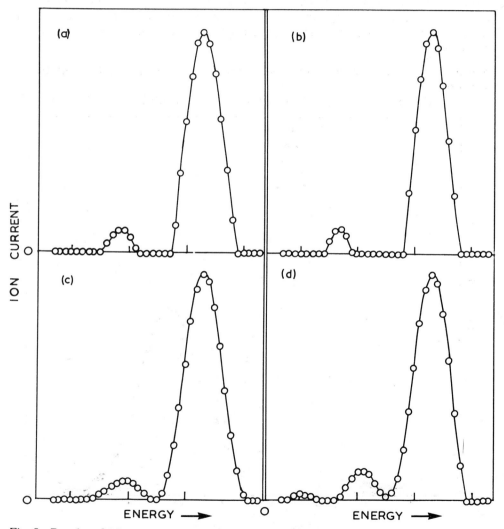

Fig. 8: Results of 15 smoothing iterations and 20 unfolding iterations of the noiseless convoluted form of function A using the electron energy distribution shown in Fig. 6. (a) correct, (b) overestimate, (c) underestimate and (d) correct but with curtailed form.

The results of smoothing and deconvolution iterations using these approximations may be seen for function (A) in Fig. 8. As might be expected, distribution 2 causes the peak width to be narrower than in the original data and the onsets are also displaced slightly. In the case of distribution 3 the peaks are not completely resolved and again the onsets are displaced, moving to lower energies. Distribution 4 has the most fundamental effect by producing an extra peak at lower energy, giving incomplete peak separation and poor recovery of the onset energies.

It is clear that even poor estimates of the electron energy distribution may be used to give qualitative or semi-quantitative information regarding negative ion processes. However, to obtain reliable values for thresholds, peak maxima and peak widths, a fairly accurate value of the electron energy distribution must be known. On the other hand a certain level of noise may be tolerated without prejudice to the final results if a smoothing procedure is used.

Application of the method to actual systems: O⁻ ion formation by sulphur dioxide

Negative ion formation in sulphur dioxide has been studied by several workers[3,5,17]

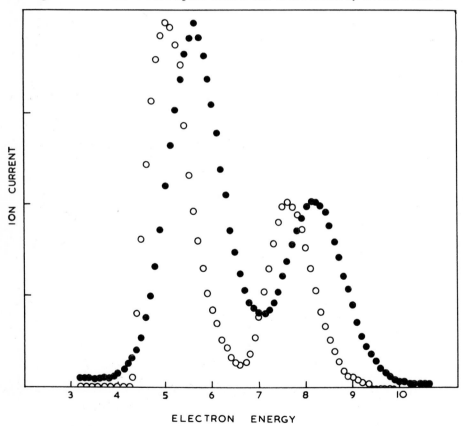

Fig. 9: Ionisation efficiency curves for O⁻ ion formation by sulphur dioxide. Full circles: original experimental data; open circles: deconvoluted results. 15 smoothing and 16 unfolding iterations. Electron energy distribution shown in Figure 7a used.

and the appearance potential of the O⁻ ion is sufficiently well established at 4·2 eV to be used to calibrate the energy scale. We have used this ion to evaluate our energy scale because it appears in a region close to where most of the measurements on the other systems were made.

Our experimental data for O^-/SO_2 are shown in Fig. 9; two resonance peaks are noted which have uncertain onsets but clear maxima. Deconvolution of these data lead to the peaks shown by the open circles; the threshold energy for the first peak is sharp and is separated by 0·80 eV from the resonance maximum. This is in exact agreement with the difference obtained by Kraus[3] and by Dillard and Franklin.[5]

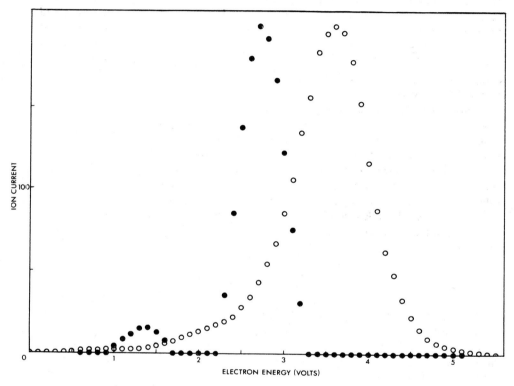

Fig.10: Ionisation efficiency curve for F⁻ ion formation by carbonyl fluoride. Open circles indicate original data; closed circles the results of 15 smoothing and 16 unfolding iterations using the electron energy distribution shown in Figure 7a.

The onset for the second resonance peak is not quite resolved, the minimum depth being 6% of the height of the first peak. Kraus,[3] using a retarding potential difference method, was also not able to completely separate the two peaks. Measurements on Fig. 10 in his paper indicate the minimum depth/peak height ratio to be 0·07, in very good agreement with the figure we report. We noted that, in this case, increasing the number of deconvolution iterations did not improve the peak separation. The energy difference between the onset of the first resonance peak and the minimum was 2·4 eV for both our work and that of Kraus. The energy difference between the maxima of the two peaks is

2·6 eV; Kraus finds 3·0 eV. Both resonance peaks have the same width at half-peak-height (1·0 eV).

Application of the method to actual systems: Ion formation by carbonyl fluoride

Negative ion formation occurred extensively in carbonyl fluoride, the most abundant ion being F^-. At 70 eV the relative intensities of F^- : F_2^- : CFO^- were 1000:45:6 and at the respective resonance maxima, 1000:104:30.

Our basic experimental data and the unfolded results are shown in Figs. 10-12 and Table 1. At about 0·71 eV it is apparent that there are very low intensity peaks for all three ions; these peaks were always observed in this region but their intensity appeared to depend upon the filament temperature. In order to maintain a constant electron current at very low energies ($\sim$1 eV) in our experimental set-up, increase in the filament temperature is required. We believe that these low intensity ions originate from the thermal decomposition of carbonyl fluoride on the filament, and at low energies (and hence high filament temperatures) this results in a small amount of ion formation. In this respect we have observed[11] the formation of polysulphur negative ions (up to S_6^-) in carbonyl sulphide and in carbon disulphide at the same low electron energy ($\sim$1 eV), under source conditions where consecutive ion-molecule reactions[5] could not occur. These ions also appear to be formed subsequent to thermal decomposition of the sulphides on the filament. A further reason for rejecting these peaks as spurious to the 'real' ionisation processes is that energetic calculations based on their appearance yield quite absurd conclusions, whereas values obtained using the major resonance peaks lead to reasonable results.

F^- ion formation. Possible dissociative capture processes which could account for ion formation in the F^- region are:

$$CF_2O + e \rightarrow F^- + CFO \quad (4)$$
$$\rightarrow F^- + F + CO \quad (5)$$

Our data (Fig. 10) indicate a single resonance peak having a sharp threshold at $2·1 \pm 0·1$ eV (average of 7 runs), the peak reaching a maximum at $2·7 \pm 0·1$ eV and having a width at half-height of $0·65 \pm 0·05$ eV. The enthalpy of formation of the fluoroformyl radical is not known, so that the enthalpy requirements for reaction (4), ΔH_4, cannot be deduced, however, $\Delta H_5 \geqslant 3·6$ eV. It is therefore apparent that reaction (4) is responsible for ion formation at 2·1 eV, and that no process corresponding to reaction (5) is occurring, since the ion current falls to zero above 3·3 eV.

Table 1. Appearance potentials (AP), peak maxima (M) and peak widths at half-height (PW) of negative ions by carbonyl fluoride. (All values in electron volts).

Ion	AP	M	PW	Process
F^-	$2·1 \pm 0·1$	$2·7 \pm 0·1$	$0·65 \pm 0·05$	$CF_2O + e \rightarrow F^- + CFO$
F_2^-	$2·6 \pm 0·1$	$3·1 \pm 0·1$	$0·65 \pm 0·04$	$\rightarrow F_2^- + CO$
CFO^-	$2·8 \pm 0·1$	$3·3 \pm 0·1$	$0·54 \pm 0·06$	$\rightarrow CFO^- + F$

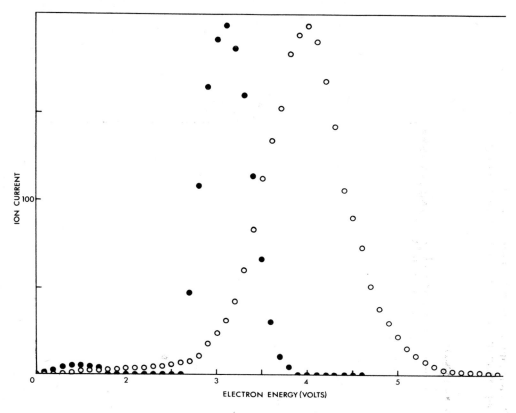

Fig.11: Ionisation curve for F_2^- ion formation by carbonyl fluoride; open circles: experimental data; closed circles: deconvoluted results.

Using the relation:

$$D(F - CFO) \leqslant A(F^-) + E(F)$$

in conjunction with the value[12] of 3·4 eV for $E(F)$ we find that $D(F - CFO) \leqslant 5·5$ eV. This may be compared with reported C-F bond dissociation energies of 5·3 eV (CF_4),[13] 5·0 eV (C_6H_5F)[14] and 5·2 eV (C_2F_6).[15]

$D(F-CFO) + D(F-CO)$ may be calculated to be 6·9 eV, so that $D(F-CO) \geqslant 1·4$ eV and hence $\Delta H_f(FCO) \sim -1·7$ eV. This bond dissociation energy may be compared with the value of about 0·4 eV for the corresponding bond in the formyl radical.[16]

F_2^- ion formation. We examined the pressure dependence of the F_2^- ion current at the peak maximum and found it to be linear; it is therefore apparent that this ion is formed by a rearrangement process and not by an ion-molecule reaction.

$$CF_2O + e \rightarrow F_2^- + CO \quad (6)$$

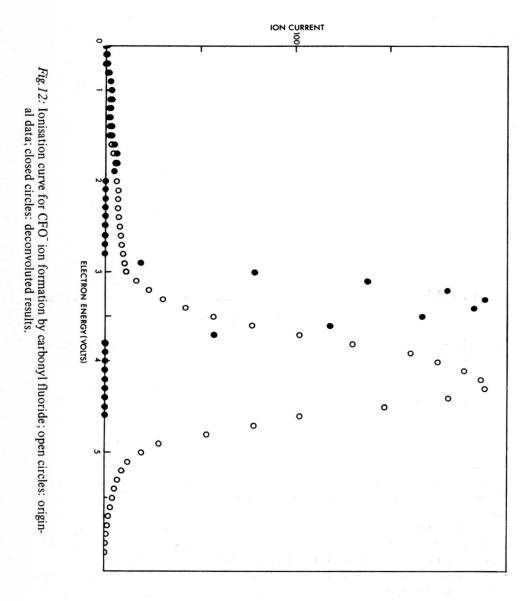

Fig.12: Ionisation curve for CFO⁻ ion formation by carbonyl fluoride; open circles: original data; closed circles: deconvoluted results.

Our basic and unfolded data are shown in Fig. 11; the ion has an onset potential of 2.6 ± 0.1 eV (average of 8 runs) and the resonance peak is quite symmetrical; the peak width at half-height is 0.65 ± 0.04.

The F_2^- ion is observed in the negative ion mass spectrum of many fluorinated compounds which suggests that it has quite a large electron affinity. Since it is clearly stable with respect to the decomposition $F_2^- \rightarrow F^- + F$, then its electron affinity must be greater than 1.8 eV. A value of 2.8 eV for $E(F_2)$ can be estimated from our data; this is in good agreement with the value of 3.0 eV obtained from F_2^-/SO_2F_2.[17]

CFO$^-$ ion formation. A symmetrically-shaped resonance peak is observed (see Fig. 12) for the CFO$^-$ ion, the appearance potential being 2.8 ± 0.1 eV (average of 5 runs), and the half-height peak width 0.54 ± 0.06 eV. The ionisation process must be:

$$CF_2O + e \rightarrow CFO^- + F \qquad (7)$$

If no excitation or kinetic energy is associated with reactions (4) and (7), then:

$$A(CFO^-) + E(CFO) = A(F^-) + E(F)$$

Our data therefore indicate a value of 2.7 eV for the electron affinity of the fluoroformyl radical. We know of no other value with which our results may be compared.

Applications of the method to actual systems: Hexafluoroacetone (HFA)

In Table 2 we show the negative ion mass spectrum of HFA measured at nominal electron energies of 10 and 70 eV. The absence of ions such as C^-, O^- and C_2^- from the lower energy spectrum suggests that they are formed principally by ion-pair processes. We have studied the energy dependence of formation of the following ions: $CF_3COCF_3^-$, $CF_3COCF_2^-$, $(CF_3)_2C^-$, CF_3CO^-, CF_3^-, CF_2^-, CFO^- and F^-.

It is apparent from our data (discussed below) that several of the ions ($CF_3COCF_2^-$, CF_3CO^-, CF_3^-, CFO^- and F^-) have almost identical appearance potentials ($\sim 3.1 \pm 0.1$ eV) and their respective resonance peaks attain a maximum value at 4.2 ± 0.1 eV. This suggests a common origin for these ions and we suggest that this is an electronically excited unstable state of the ketone which subsequently decomposes to form the ion mentioned:

$$CF_3COCF_3 + e \rightarrow (CF_3COCF_3^-)^* \rightarrow CF_3COCF_2^- + F \qquad (8)$$
$$\rightarrow CF_3CO^- + CF_3 \qquad (9)$$
$$\rightarrow CF_3^- + COCF_3 (+ CO + CF_3?) \qquad (10)$$
$$\rightarrow CFO^- + CF_2 + CF_3 \qquad (11)$$
$$\rightarrow F^- + CF_2COCF_3 \qquad (12)$$

Charge-transfer reactions involving the O^- ion. The ionisation curves for several of the ions (e.g. $CF_3COCF_2^-$, CF_3^-, F^-) show inflections at energies ~ 5 eV; this corresponds to the

Table 2. Negative ion mass spectrum of hexafluoroacetone at electron energies (uncorrected) of 10 eV and 70 eV.

m/e	Ion	Rel. Int. (10 eV)	Rel. Int. (70 eV)
12	C^-	0	3.0
16	O^-	0	21.5
19	F^-	751	1000
24	C_2^-	0	4.0
31	CF^-	15.2	3.6
38	F_2^-	0	3.0
43	C_2F^-	0	2.0
47	CFO^-	25.4	18.3
50	CF_2^-	12.7	4.0
69	CF_3^-	1000	257
97	CF_3CO^-	10.1	17.0
147	$CF_3COCF_2^-$	5.7	36.0
150	$(CF_3)_2C^-$	7.6	4.0
166	$CF_3COCF_3^-$	7.6	67

resonance peak maximum observed for the O^- ion formed by sulphur dioxide (which was used to calibrate the electron energy scale). These inflections were not observed when sulphur dioxide was not present in the ionisation chamber; accordingly we attribute their occurrance to the charge-transfer reaction (13). This produces an unstable state of the parent ion which decomposes to yield the appropriate ions.

$$O^- + CF_3COCF_3 \rightarrow O + CF_3COCF_3^- \rightarrow CF_3COCF_2^- \text{ etc.} \qquad (13)$$

$CF_3COCF_3^-$ ion formation. The observation of a stable molecule-ion such as $CF_3COCF_3^-$ is relatively unusual, few other such ions having been observed. Its formation must involve the electron capture process:

$$CF_3COCF_3 + e \rightarrow CF_3COCF_3^- \qquad (14)$$

Our data for this ion indicates two main regions of ion formation, one being at ~ 0 eV and the second commencing at about 10 eV. In earlier work[18] only the higher energy process was detected, the ion being formed as a result of secondary electron capture, the secondary electrons being produced by such positive ionisation processes as:

$$CF_3COCF_3 + e \rightarrow CF_3COCF_3^+ + 2e$$

At that time we were unable to observe ion formation at very low electron energies; however, an improved experimental technique and introducing a 3 V dry cell into the electron energy circuit so that 'negative' voltages could be obtained has enabled us to examine the primary electron capture reaction.

When hexafluoroacetone was studied at electron energies $\sim$0 eV, a parent ion was observed; admission of a small quantity of sulphur hexafluoride to the ionisation chamber resulted in a considerable reduction in the $CF_3COCF_3^-$ ion intensity. This suggested that the electron attachment cross-section for reaction (14) was much less than that for SF_6^- ion formation by reaction (15)

$$SF_6 + e \rightarrow SF_6^- \qquad (15)$$

In Fig. 13 we show the data obtained for SF_6^- and $CF_3COCF_3^-$ ion formation as a function of the electron energy. A 50/50 mixture of hexafluoroacetone and sulphur hexafluoride was used; the two sets of ionisation data have been normalised for convenience in presentation, the ordinate for $CF_3COCF_3^-$ being 58·9 times greater than that for SF_6^-.

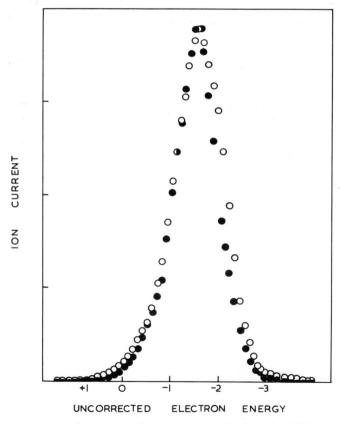

Fig. 13: Ion current versus electron accelerating energy. Full circles SF_6^-, open circles $CF_3COCF_3^-$. Ion current scale for hexafluoroacetone 58·9 times greater than that for SF_6.

Formation of the SF_6^- ion in this energy region has been used to mirror the electron energy distribution and calibrate the electron energy scale.[6,7] It is apparent from Fig. 13 that the $CF_3COCF_3^-$ and SF_6^- ions have a very similar energy dependence, both ions attaining a maximum value at the same electron energy. The ionisation curve for the ketone is slightly broader in the wings than that for the hexafluoride; this may reflect either a slightly different energy dependence for electron attachment or the experimental uncertainties in measuring the very small ion currents for the $CF_3COCF_3^-$.

Because of their similar energy dependence we consider that the relative heights of the two ion peaks may be used to indicate the relative attachment cross-sections of reactions (14) and (15). If it is assumed that both ions have the same collection efficiency, then:

$$\frac{\sigma_{SF_6}}{\sigma_{HFA}} = 58.9$$

where σ_X refers to the electron attachment cross-section of X. Using an electron-swarm technique Compton et al.[19] have obtained a value of 3.6×10^{-15} cm^2 for the electron attachment cross-section for SF_6. Using this result in conjunction with the ratio reported above we find that $\sigma_{HFA} = 0.61 \times 10^{-16}$ cm^2.

It has been observed that both sulphur hexafluoride[20] and hexafluoroacetone[18] form parent molecule-ions at higher electron energies as a result of secondary electron capture. We considered that competition between sulphur hexafluoride and hexafluoroacetone for secondary electrons might enable us to measure $\sigma_{SF_6}/\sigma_{HFA}$. Accordingly, using a 39.2:1 mixture of $CF_3COCF_3:SF_6$, we measured the intensities of the $CF_3COCF_3^-$ and SF_6^- ions, I_{HFA^-} and $I_{SF_6^-}$, at ten electron energy intervals over the range 15–60 eV. Our experimental data indicated that the ion current ratio $I_{SF_6^-}/I_{HFA^-}$ was effectively constant over the entire energy range having a value of 1.44 ± 0.06; this result yields a value for $\sigma_{SF_6}/\sigma_{HFA}$ of 56 ± 2, which is in good accord with our directly measured value at low electron energies.

$CF_3COCF_2^-$ ion formation. Typical experimental data for $CF_3COCF_2^-$ and the smoothed, deconvoluted results are shown in Figs. 14 and 15 together with the corresponding O^-/SO_2 ionisation curves.

A sharp onset at 3.10 ± 0.10 eV is observed, the resonance peak reaching a maximum at 4.20 ± 0.05 eV and the peak width at half-height is 1.35 ± 0.05 eV. The common origin of the ions formed at this energy has been discussed above and we attribute $CF_3COCF_2^-$ ion formation to reaction (8).

If a value of ~ 5.2 eV is assumed for the bond dissociation energy $D(CF_3COCF_2-F)$, (values of 5.3, 5.2 and 5.0 eV having been reported for the C–F bond strengths in CF_4,[13] C_6H_5F[14] and C_2F_6[15] respectively), then using the relation:

$$D(CF_3COCF_2-F) \leq A(CF_3COCF_2^-) + E(CF_3COCF_2),$$

a value of ~ 2.1 eV may be estimated for the electron affinity of the perfluoroacetonyl radical.

Figure 15 shows clearly the inflection in the ionisation curve which we have considered above to be the results of the charge-transfer reaction (13).

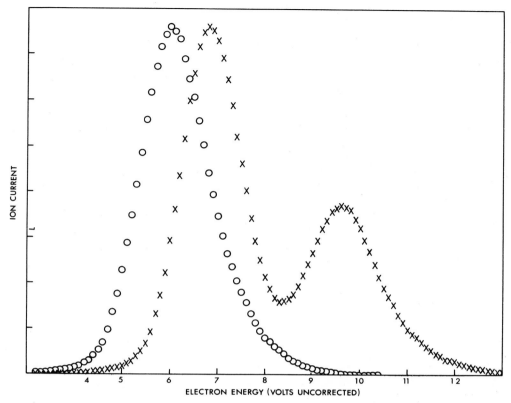

Fig.14: Ionisation efficiency curve for $CF_3COCF_2^-/CF_3COCF_3$ (o) and O^-/SO_2 (x) ion formation.

$(CF_3)_2C^-$ ion formation. Our experimental data for the $(CF_3)_2C^-$ ion when smoothed and deconvoluted yield the values shown in Table 3. Ion formation is attributed to the reaction:

$$CF_3COCF_3 + e \rightarrow (CF_3)_2C^- + O$$

It is noteworthy that the resonance peak is very narrow, the width at half-height being only 0.6 eV. A value of ~ 0.6 eV may be estimated for the electron affinity of $(CF_3)_2C$ if the bond strength $D(O-C(CF_3)_2)$ is assumed to be similar to that in carbon dioxide, i.e. ~ 5.7 eV.

CF_3CO^- ion formation. At low electron energies the CF_3CO^- ion is formed quite abundantly but not at 70 eV; this suggests that the ion-pair process:

$$CF_3COCF_3 + e \rightarrow CF_3CO^- + CF_3^+ + 2e$$

does not occur extensively.

Our results for this ion are shown in Table 3, and we consider reaction (9) to account for ion formation.

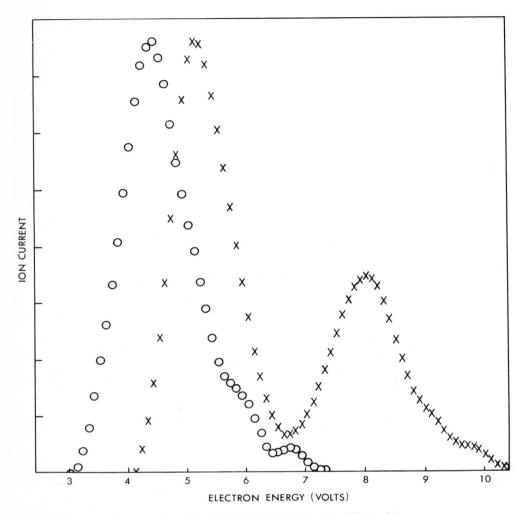

Fig. 15: Deconvoluted results: $CF_3COCF_2^-/CF_3COCF_3$ (o); O^-/SO_2 (x).

If a value of 3·7 eV is assumed for $D(CF_3CO\text{-}CF_3)$, then our data yields a value of 0·6 eV for the electron affinity of the trifluoroacetyl radical.

CF_3^- *ion formation.* Ion formation in the case of CF_3^- is rather complex, several appearance potentials being noted for this ion.

Initially ion formation occurred at 3·0 ± 0·1 eV and is attributed to decomposition of the electronically excited ketone by reaction (10). Because the trifluoroacetyl radical has been shown to have limited stability in the gas phase, the decomposition reaction (10) may involve the formation of CO and CF_3 as fragmentation products.

It is apparent from Fig. 17 that the CF_3^- ionisation curve has inflections at ∿ 5 eV and ∿ 7·6 eV, i.e. where the O^-/SO_2 ion reaches a maximum intensity; these inflections are attributed to the charge-transfer reaction mentioned above.

Table 3. Appearance potentials (AP), peak maxima (M) and peak widths at half-height (PW) of negative ions formed by hexafluoroacetone.

Ion	AP	M	PW	Process
$CF_3COCF_3^-$	0			$CF_3COCF_3 + e \rightarrow CF_3COCF_3^-$
$CF_3COCF_2^-$	3·10 ± 0·10	4·20 ± 0·05	1·35 ± 0·05	$CF_3COCF_3 + e \rightarrow (CF_3COCF_3^-)^*$
				$\downarrow$
				$CF_3COCF_2^- + F$
$(CF_3)_2C^-$	5·1 ± 0·1	6·4 ± 0·1	0·6 ± 0·1	$CF_3COCF_3 + e \rightarrow (CF_3)_2C^- + O$
CF_3CO^-	3·10 ± 0·05	4·1 ± 0·1	1·0 ± 0·2	$(CF_3COCF_3^-)^* \rightarrow CF_3CO^- + CF_3$
CF_3^-	3·0 ± 0·1	4·3 ± 0·1	1·5 ± 0·2	$(CF_3COCF_3^-)^* \rightarrow CF_3^- + CF_3CO$
	5·4 ± 0·1	6·7 ± 0·1	1·7 ± 0·1	$CF_3COCF_3 + e \rightarrow CF_3^- + CO + CF_2 + F$
	8·2 ± 0·2	8·6 ± 0·1	–	?
	10·5 ± 0·2	11·0 ± 0·2	–	?
CF_2^-	4·25 ± 0·10	5·25 ± 0·05	0·7 ± 0·1	$CF_3COCF_3 + e \rightarrow CF_2^- + F + CO + CF_3$?
CFO^-	3·0 ± 0·1	uncertain	~2	$(CF_3COCF_3^-)^* \rightarrow CFO^- + CF_2 + CF_3$
	5·3 ± 0·2	6·6 ± 0·1	2·0 ± 0·1	$CF_3COCF_3 + e \rightarrow CFO^- + 2F + C_2F_3$
F^-	3·1 ± 0·1	4·3 ± 0·2	1·3 ± 0·3	$(CF_3COCF_3^-)^* \rightarrow F^- + CF_2COCF_3$
	5·7 ± 0·1	7·2 ± 0·2	1·8 ± 0·3	$CF_3COCF_3 + e \rightarrow F^- + CF_2 + CO + CF_3$
	9·0 ± 0·1	–	–	$\rightarrow F^- + F + CO + 2CF_2$

A second resonance process is observed at 5·4 eV (based upon extrapolation of the upper part of the curve). If reaction (16) is responsible for ion formation at this energy, then our data yield a value of 2·5 eV for $E(CF_3)$; this may be compared with values of 2·6 eV[15] and 1·8 eV[21] reported for this quantity.

$$CF_3COCF_3 + e \rightarrow CF_3^- + CO + CF_2 + F \tag{16}$$

Further resonance processes of very low cross-section are noted at 8·2 and 10·5 eV; we cannot account for these ionisation processes.

CF_2^- ion formation. Our data for the CF_2^- ion are shown in Table 3; a narrow resonance peak having an onset at 4·25 ± 0·10 eV is obtained. If ion formation was due to the reaction:

$$CF_3COCF_3 + e \rightarrow CF_2^- + F + CO + CF_3 \tag{17}$$

then a value of $E(CF_2) \leqslant 3·75$ eV would be obtained; this seems improbably large. The narrow resonance peak would suggest that little excess energy was involved in the ionisation process. A rearrangement reaction such as:

$$CF_3COCF_3 + e \rightarrow CF_2^- + CF_4 + CO \tag{18}$$

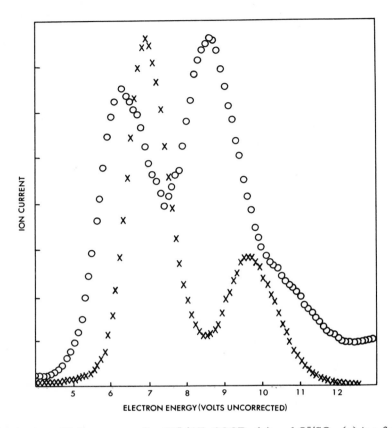

Fig. 16: Ionisation efficiency curve for CF_3^-/CF_3COCF_3 (o) and O^-/SO_2 (x) ion formation.

would yield a negative value for $E(CF_2)$; we are therefore unable to assign the ionisation process resonsible for CF_2^- formation.

CFO$^-$ ion formation. The ion CFO$^-$ must be formed as the result of rearrangement; ion formation is observed initially at $3 \cdot 0 \pm 0 \cdot 1$ eV and is attributed to reaction (11). Although a sharp onset is obtained at this energy the resonance peak is broad (~ 2 eV) and does not attain a clear maximum before a second resonance process occurs at 5·3 eV. This resonance peak is also broad and this perhaps suggests the involvement of considerable excess energy in the rearrangement.

$$CF_3COCF_3 + e \rightarrow CFO^- + 2F + C_2F_3 \tag{19}$$

The second ionisation process may correspond to reaction (19), i.e. the C-O bond is broken and the oxygen transferred to a CF group in the rearrangement. This probably would have higher energy requirements than reaction (11) where simple fluoride transfer to CO may be involved, but such assignments are tentative.

The heat of formation of the fluoroformyl radical, ΔH_f (CFO), has been estimated above from data on carbonyl fluoride to be $-1 \cdot 7$ eV, so that our data for reaction (11) indicate E(CFO) $\sim 3 \cdot 3$ eV. Our investigation of CFO$^-$ ion formation by carbonyl fluoride showed E(CFO) = $2 \cdot 7$ eV; the resonance peak in that study is much narrower than that

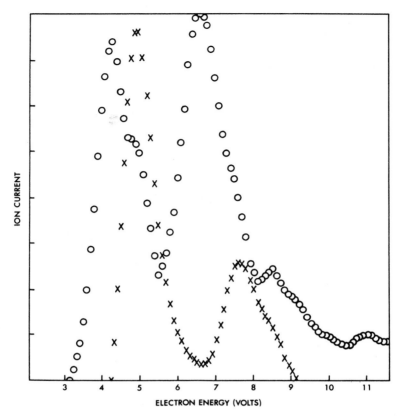

Fig. 17: Deconvoluted results: CF_3^-/CF_3COCF_3 (o); O^-/SO_2 (x).

found in this work (~ 0.5 eV compared with ~ 2.0 eV), which may reflect the excess energy involved in the hexafluoroacetone rearrangement.

Our results for the F^- ion are summarised in Table 3 and typical data shown in Figs. 18 and 19.

Inflections in the deconvoluted curve at ~ 5 and ~ 7.5 eV we attribute to the charge-transfer reaction involving the O^- ion. The first appearance potential at 3.1 ± 0.1 eV is considered to be the result of reaction (12) discussed previously; a second process of much larger cross-section occurs at 5.7 ± 0.1 eV. If reaction (20) is responsible for the increase in the ion current at this energy then a maximum value of 2.6 eV may be deduced for the bond dissociation energy $D(CF_2-COCF_3)$ if we assume that $D(CO-CF_3) = 0$ eV.

$$CF_3COCF_3 + e \rightarrow F^- + CF_2 + CO + CF_3 \qquad (20)$$

A further ionisation process of very low cross-section may be seen on the tail of the second resonance peak, the onset energy being 9.0 eV.

$$CF_3COCF_3 + e \rightarrow F^- + F + CO + 2CF_2 \qquad (21)$$

The minimum enthalpy requirements for reaction (21) are 9.0 eV and we therefore attribute ionisation to this reaction.

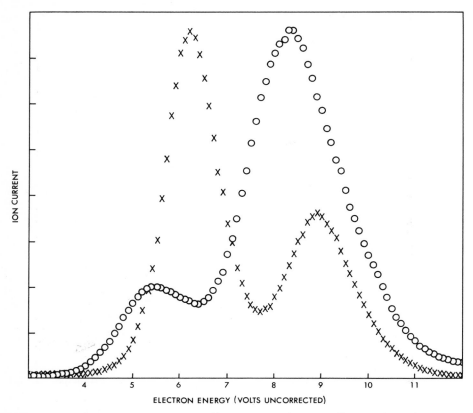

Fig. 18: Ionisation efficiency curve for F^-/CF_3COCF_3 (o) and O^-/SO_2 (x) ion formation.

Application of the method to actual systems: Negative ion formation by sulphur hexafluoride

The negative ions observed at 70 eV are F^-, S^-, F_2^-, SF^-, SF_2^-, SF_3^-, SF_4^-, SF_5^- and SF_6^-, with F^- and SF_6^- being the most abundant. The formation of most of these ions is well-known, in particular the SF_6^- ion which has been used to determine the electron energy distribution.[6,7]

SF_5^- ion formation. The SF_5^- ion is formed abundantly, presumably by the dissociative capture reaction:

$$SF_6 + e \rightarrow SF_5^- + F$$

and we have measured an appearance potential of 0·1 eV for this ion, a result in good accord with those of other workers.[20,22]

SF_4^- ion formation.

$$SF_6 + e \rightarrow SF_4^- + 2F \tag{22}$$

Our experimental data indicate that $A(SF_4^-) = 5·0 \pm 0·1$ eV; if reaction (22) is

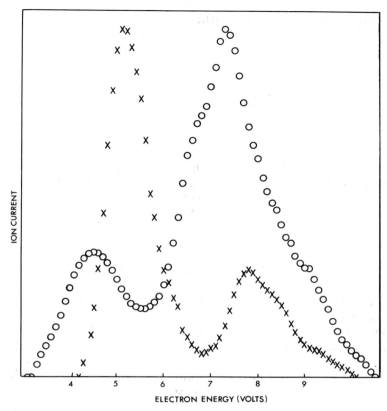

Fig. 19: Deconvoluted results: F^-/CF_3COCF_3 (o); O^-/SO_2 (x).

responsible for ion formation then a value of 1·7 eV may be estimated for the electron affinity of sulphur tetrafluoride, $E(SF_4)$.

Table 4. Appearance potentials (A), resonance peak maxima (M) and half-widths (PW) for negative ions formed by SF_6. (All values in electron volts)

Ion	A	M	PW	Process
SF_6^-	0	0·4 ± 0·1	0·6 ± 0·1	$SF_6 + e \to SF_6^-$
SF_5^-	0·1	0·5 ± 0·1	0·8 ± 0·1	$\to SF_5^- + F$
SF_4^-	5·0 ± 0·1	6·0 ± 0·1	1·7 ± 0·1	$\to SF_4^- + 2F$
F^-	4·3 ± 0·1	5·7 ± 0·1	1·4 ± 0·1	$\to SF_4 + F + F^-$
	7·8 ± 0·1	9·3 ± 0·1	~2	$\to SF_3 + 2F + F^-$
	10·5 ± 0·1	11·8 ± 0·1	1·6 ± 0·1	$\to SF_2 + 3F + F^-$

F⁻ ion formation was observed initially at electron energies near to zero, indicating that $D(SF_5-F) \sim E(F)$. Three other ionisation processes were noted at 4·3, 7·8 and 10·5 eV, the process 4·3 eV having considerably the largest cross-section.

$$SF_6 + e \rightarrow SF_4 + F + F^- \qquad (23)$$

$$\rightarrow SF_3 + 2F + F^- \qquad (24)$$

$$\rightarrow SF_2 + 3F + F^- \qquad (25)$$

The minimum enthalpy requirements for reaction (23) are 3·3 eV, so we suggest that this reaction is responsible for the increase in ion current at 4·3 eV. Reactions (24) and (25) have higher energy requirements than (23), in both cases S-F bonds being broken which probably require $\sim$ 3 eV. The energy difference observed between the ionisation processes are 3·5 and 2·7 eV; this is close to the likely S-F bond energies and we tentatively suggest that reactions (24) and (25) also participate in ion formation.

Applications of the method to actual systems: Negative ion formation by SF₅Cl

The negative ion mass spectrum measured at (uncorrected) electron energies of 2 and 70 eV are shown in Table 5. F⁻ is the most abundant ion but, unlike SF_6, no parent negative ion is observed either at low electron energies (where it could be formed by primary electron capture) or at higher energies where secondary electrons might be involved in the capture process. No ions which contain S, F and Cl or S and Cl are formed. Some Cl⁻ and F⁻ ion formation occurs at $\sim$ 0 eV as a consequence of thermal decomposition of the pentafluorosulphur chloride on the hot filament.

Table 5. Negative ion mass spectrum of SF_5 Cl at 2 and 70 eV

m/e	Ion	Rel. Int. (2)	Rel. Int. (70)
19	F⁻	1000	1000
32	S⁻	–	0·65
35	Cl⁻	26·7	100
38	F_2^-	6·7	–
51	SF⁻	2·2	8·0
54	FCl⁻	2·2	2·0
70	SF_2^-	2·0	0·65
89	SF_3^-	2·2	16·2
108	SF_4^-	2·3	3·3
127	SF_5^-	20·0	24·1

SF_5^- ion formation. The SF_5^- is observed at $0·2 \pm 0·1$ eV, the resonance peak attaining a maximum value of 0·7 eV. Ionisation is attributed to the reaction:

$$SF_5Cl + e \rightarrow SF_5^- + Cl \qquad (26)$$

SF_4^- ion formation. Our appearance potential data for the SF_4^- ion are given in Table 6.

$$SF_5Cl + e \rightarrow SF_4^- + F + Cl \qquad (27)$$

If reaction (27) is responsible for ion formation then we can estimate a value of 1·0 eV for $E(SF_4)$; this may be compared with the value of 1·7 eV deduced for $E(SF_4)$ in the experiments using SF_6.

Table 6. Appearance potentials (A), resonance peak maxima (M) and half-widths (PW) for negative ions formed by pentafluorosulphur chloride. (All values in electron volts)

Ion	A	M	PW	Process
SF_5^-	0·2 ± 0·1	0·70 ± 0·10	0·8 ± 0·1	$SF_5Cl \rightarrow SF_5^- + Cl$
SF_4^-	4·1 ± 0·1	4·65 ± 0·1	1·0 ± 0·1	$\rightarrow SF_4^- + F + Cl$
SF_3^-	7·9 ± 0·2	8·9 ± 0·1	1·4 ± 0·2	$\rightarrow SF_3^- + 2F + Cl$
FCl^-	7·6 ± 0·1	9·1 ± 0·1	1·7 ± 0·1	$\rightarrow SF_2 + 2F + FCl^-$
Cl^-	4·0 ± 0·1	uncertain	—	$\rightarrow SF_3 + F_2 + Cl^-$
	7·6 ± 0·2	9·1 ± 0·1	1·6 ± 0·1	$\rightarrow SF_2 + 3F + Cl^-$?
F^-	3·2 ± 0·1	5·1 ± 0·1	1·3 ± 0·1	$\rightarrow SF_3 + FCl + F^-$
	6·3 ± 0·2	6·5 ± 0·1	—	$\rightarrow SF_2 + F_2 + Cl + F^-$
	8·0 ± 0·1	9·4 ± 0·1	∼1·5	$\rightarrow SF_2 + 2F + Cl + F^-$

SF_3^- ion formation. The dissociation into atoms of sulphur tetrafluoride requires 13·4 eV; if we assume that the four S-F bonds have equal strengths then $D(SF_3-F) \sim 3·35$ eV and we may estimate the heat of formation of SF_3, $\Delta H(SF_3) = 4·9$ eV.

$$SF_5Cl + e \rightarrow SF_3^- + 2F + Cl \qquad (28)$$

$A(SF_3^-) = 7·9$ eV so that, from reaction (28), we may estimate the electron affinity of SF_3 to be 0·6 eV. The similarity of the appearance potentials of SF_3^- and FCl^- initially suggest the occurrence of the reaction:

$$SF_5Cl + e \rightarrow SF_3^- + F + FCl$$

but the energetic requirements for this reaction are such as to indicate $E(SF_3) < 0$, and so the reaction is neglected.

FCl^- ion formation. The FCl^- ion must be formed either by an ion-molecule reaction involving F^- or Cl^-, e.g.

$$F^- + SF_5Cl \rightarrow FCl^- + SF_5, \qquad (29)$$

or by rearrangement reactions such as:

$$SF_5Cl + e \rightarrow SF_4 + FCl^- \tag{30}$$
$$\rightarrow SF_3 + F + FCl^- \tag{31}$$
$$\rightarrow SF_2 + 2F + FCl^- \tag{32}$$

A study of the pressure dependence of FCl^- ion formation at 7.6 eV showed it to be a primary ion and we may therefore rule out secondary reactions such as (29).

Reactions (30) and (31) may also be neglected as sources of the FCl^- ion on energetic grounds. To estimate the energetics of (32) we require to know $\Delta H(SF_2)$. If we assume $D(SF_2-F) \sim 3.3$ eV, then $\Delta H(SF_2) \sim -2.4$ eV. Using this estimate we can calculate that $E(FCl) \sim 1.6$ eV. We know of no value with which this may be compared but values of 2.8 eV and $\leqslant 1.7$ eV have been obtained for $E(F_2)$ (see above) and for $E(Cl_2)$[23] respectively.

Cl⁻ ion formation. Our data for the Cl^- ion show it to have an appearance potential at 4.0 eV, the broad resonance peak rising slowly with a much more intense resonance process having an appearance potential at 7.6 ± 0.2 eV, the peak maximum being attained at 9.1 eV. Some ion formation also occurred at electron energies ~ 0 eV; this may be due to thermal decomposition of SF_5Cl or to the occurrence of reaction (33) since $D(SF_5-Cl) < E(Cl)$.

$$SF_5Cl + e \rightarrow SF_5 + Cl^- \tag{33}$$
$$\rightarrow SF_4 + F + Cl^- \tag{34}$$
$$\rightarrow SF_3 + 2F + Cl^- \tag{35}$$
$$\rightarrow SF_3 + F_2 + Cl^- \tag{36}$$
$$\rightarrow SF_2 + 3F + Cl^- \tag{37}$$

The broad peak which has its onset at 4.0 eV may be attributable to reaction (36) for which the enthalpy requirements are ~ 3.4 eV. Reactions (35) and (36) would correspond to appearance potenials for Cl^- of 4.9 and 8.2 eV; we are therefore unable to attribute the ionisation process more than tentatively and suggest reaction (37) is responsible.

F⁻ ion formation. A typical ionisation efficiency curve for the F^- ion before and after performing 15 smoothing and 20 unfolding iterations is shown in Fig. 20.

Ion formation of low intensity occurs at energies near to zero, presumably due to the reaction:

$$SF_5Cl + e \rightarrow SF_4 + Cl + F^-$$

A further dissociative capture process occurs at 3.2 eV, the resonance peak reaching a maximum value at 5.1 eV. A reaction of very low cross-section occurs at 6.3 eV and a further ionisation process at 8.0 eV.

$$SF_5Cl + e \rightarrow SF_3 + FCl + F^- \tag{38}$$
$$\rightarrow SF_2 + F_2 + Cl + F^- \tag{39}$$
$$\rightarrow SF_2 + 2F + Cl + F^- \tag{40}$$

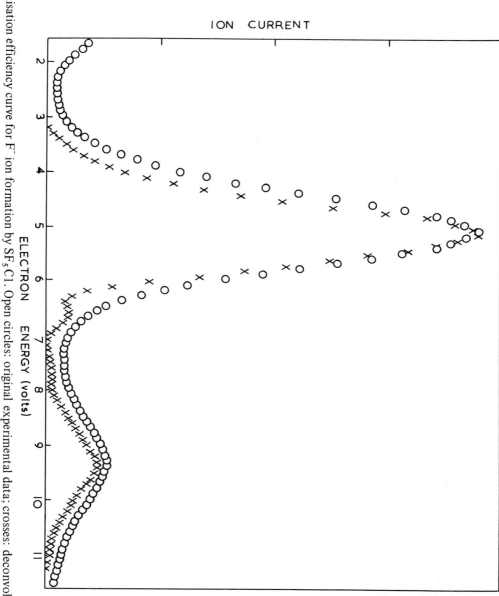

Fig. 20: Ionisation efficiency curve for F⁻ ion formation by SF_5Cl. Open circles: original experimental data; crosses: deconvoluted results.

Reaction (38) would require that $A(F^-) \sim 2.5$ eV, in reasonable accord with the onset noted at 3·2 eV. Reactions (39) and (40) have minimum enthalpy requirements of 6·8 and 8·4 eV respectively, which in view of the uncertainty associated with some of our thermochemical estimates, suggests that they may be assigned to the resonance processes occurring at 6·3 and 8·0 eV.

Thermochemical data

The following values for the heats of formation have been used in this work (in eV): CF_3COCF_3 −15·1; CF_3 −5·2;[24] CF_3^+ 3·7;[25] CF_2 −1·6;[26] CF 3·2;[27] F 0·8;[27] CO −1·1;[27] COF_2 −6·6;[27] SF_6 −12·5;[27] SF_5Cl −10·6;[28] SF_4 −7·45.[29]

The heat of formation of hexafluoroacetone has been estimated using the Additivity Rules[30] based upon the C-C and C-F bond contributions to the enthalpies of various fluorine-containing molecules.[13,39]

Acknowledgements

We thank the Science Research Council for a grant in aid of this work and Dr. H. L. Roberts of ICI (Mond Division) for a gift of the pentafluorosulphur chloride.

References

1 R. E. Fox, W. M. Hickman, T. Kjeldaas and D. J. Grove, *Phys. Rev.* **84**, 859 (1951).
2 R. E. Fox, W. M. Hickman, D. J. Grove and T. Kjeldaas, *Rev. Sci. Instr.* **12**, 1101 (1955).
3 K. Kraus, *Z. Naturforsch.* **16a**, 1378 (1961).
4 J. D. Morrison, *J. Chem. Phys.* **39**, 200 (1963).
5 J. G. Dillard and J. L. Franklin, *J. Chem. Phys.* **48**, 2349 (1968).
6 W. M. Hickman and R. E. Fox, *J. Chem. Phys.* **25**, 642 (1956).
7 G. J. Schulz, *J. Appl. Phys.* **31**, 1134 (1960).
8 P. H. van Cittert, *Z. Phys.* **69**, 298 (1931).
9 R. N. Bracewell and J. A. Roberts, *Australian J. Phys.* **7**, 615 (1954).
10 G. E. Ioup and B. S. Thomas, *J. Chem. Phys.* **46**, 3959 (1967).
11 K. A. G. MacNeil and J. C. J. Thynne, *J. Phys. Chem.* in press, 1969.
12 R. S. Berry and C. W. Rieman, *J. Chem. Phys.* **38**, 1540 (1963).
13 C. R. Patrick, *Advan. Fluorine Chem.* **2**, 18 (1961).
14 P. Smith, *J. Chem. Phys.* **29**, 681 (1958).
15 K. A. G. MacNeil and J. C. J. Thynne, *Int. J. Mass Spec. Ion Phys.* **2**, 1 (1969).
16 J. A. Kerr, *Chem. Rev.* **66**, 465 (1966).
17 R. M. Reese, V. H. Dibeler and J. L. Franklin, *J. Chem, Phys.* **29**, 880 (1958).
18 J. C. J. Thynne, *Chem. Commun.* 1075, (1968).
19 R. N. Compton, L. G. Christophorou, G. S. Hurst and P. W. Reinhardt, *J. Chem. Phys.* **45**, 4634 (1966).
20 A. J. Ahearn and N. B. Hannary, *J. Chem. Phys.* **21**, 119 (1953).
21 F. M. Page, personal communication, 1968.
22 W. M. Hickman and D. Berg, *Advan. Mass Spectrom.* 458 (1958).
23 N. S. Buchel'nikova, *Usp. Fiz. Nauk* **65**, 351 (1958).
24 B. S. Rabinovitch and J. F. Reed, *J. Chem. Phys.* **22**, 2092 (1954).
25 C. Lifschitz and F. A. Long, *J. Phys. Chem.* **69**, 3731 (1965).
26 J. R. Majer and C. R. Patrick, *Nature* **201**, 1022 (1964).

27   JANAF Thermochemical Tables, Dow Chemical Co., Midland, Mich., 1961.
28   H. F. Leach and H. L. Roberts, *J. Chem. Soc.* 4693 (1960).
29   R. D. W. Kemmitt and D. W. A. Sharp, *Advan. Fluorine Chem.* **4**, 221 (1965).
30   S. W. Benson and J. H. Buss, *J. Chem. Phys.* **29**, 546 (1958).

Discussion

Dr. P. F. Knewstubb: You report that the electronic systems of the instrument produce a rise of filament temperature when low electron energies are being used. The electron energy distribution used in deconvolution is measured for SF_6^- production at low electron energies. Is there not some question of a change in electron energy distribution as the filament temperature is changed, and what account was taken of this?

Dr. J. Thynne: Dr. Knewstubb raises a very valid point with respect to our results, and we agree that there will be a change in the electron energy distribution as the filament temperature changes so that the distribution we use for the deconvolution procedure will not exactly reflect that being obtained at higher energies. In our analysis of the artificial functions we examined the effects of variations in the energy distribution (Fig. 7) upon the recovery of peak parameters and found that small changes could be tolerated without prejudice to the final results. When we applied this procedure to sulphur dioxide we found (as shown in Fig. 9) that our data for the O^- ion were in very good agreement with those obtained by Kraus using the R.P.D. technique. It is therefore apparent that the variation in the energy distribution over the range 0 to 7 eV has little effect on the accuracy of our results. This view is supported also at electron energies $\sim$10 eV by unpublished data we have for O^- ion formation by carbon monoxide.

Dr. P. Zavitsanos: (1) Would you comment on the shape of your ion intensity versus electron energy curves for SF_6^- and SF_5^-?

(2) Do you know of any other molecules other than SF_6 to which electrons attach at electron energies $\sim$ 0 eV?

Dr. J. Thynne: With regard to the first question we find that the SF_6^- and SF_5^- resonance peaks are very similar in shape, the SF_5^- curve being displaced upwards along the electron energy scale by 0·1 eV and being marginally broader than the SF_6^- peak.

We have found in our work that, apart from hexafluoroacetone, perfluorobutene-1, tetrachloroethylene and trifluoroacetone form parent ions. There are also reports in the literature of nitrobenzene, tetranitromethane and perfluoromethylcyclohexane forming molecule ions.

Chapter 9

Ion Spectroscopy by Nanosecond Resolution Time-of-Flight Techniques

G. W. F. Pike

Physical Chemistry Laboratory, University of Oxford, England

Introduction

A general method of investigating molecular structure and chemical reaction processes is to analyse the manner of scattering of an incident stream of particles. The actual energy transformations that such a scattering process can detect depend on the amount of energy and momentum that can be transferred to or from the beam by the molecules under study. This in turn depends largely on the mass, velocity and charge of the incident beam together with the overall limitations of experimental observation.

Large areas of energy-momentum space (Fig. 1) cannot be adequately reached by existing diagnostic methods, particularly those regions pertinent to molecule-molecule and ion-molecule reactions. A beam of ions, however, having flexibility in both mass and velocity, is suitable for probing parts of this area (shaded in Fig. 1). The mass spectrometer contains such a controllable ion source and preliminary experiments on a Bendix Model 14-107 time-of-flight mass spectrometer indicate that meaningful information can be derived from the flight time analysis of scattering ions.

Apparatus

The Bendix 14-107 has been adapted for variable ion energies and to measure the relative flight times of individual ions to an accuracy better than 1 nsec. Ions of the required mass and energy, produced in the normal way, pass through a test cell inserted in front of the flight tube. The datum for timing is the ion energy pulse (Fig. 2) which after passing through a variable passive delay line is shaped and starts a voltage ramp in the time-to-amplitude converter (Harwell 2000 series). This ramp is stopped by a signal, suitably amplified and shaped, from the oscilloscope anode. The output voltage is then proportional to the time interval between initiation of the ramp and the arrival of the first ion at the spectrometer anode.* The rise-time of the ramp can be varied from 50 nsec to 1000 nsec, giving control over both time resolution and the width of spectrum to be

---

* This is not necessarily the first ion to strike the electron multiplier as these devices have a statistical variation in the output current produced from an identical ion input. Some of these are too small to activate the converter.

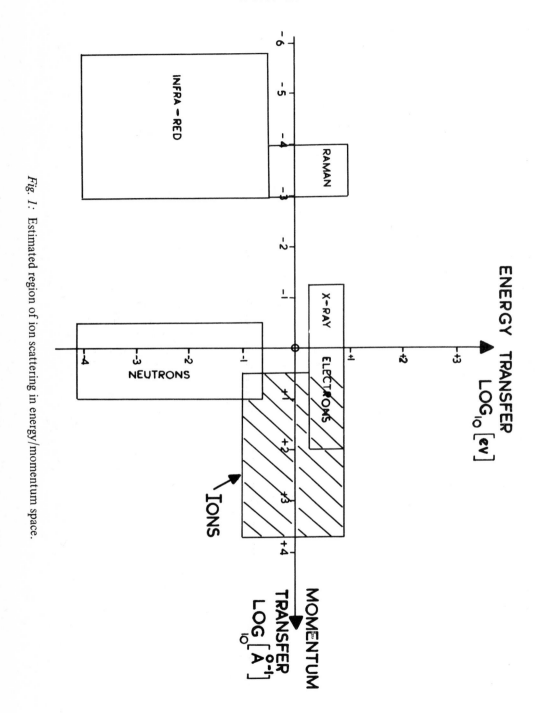

Fig. 1: Estimated region of ion scattering in energy/momentum space.

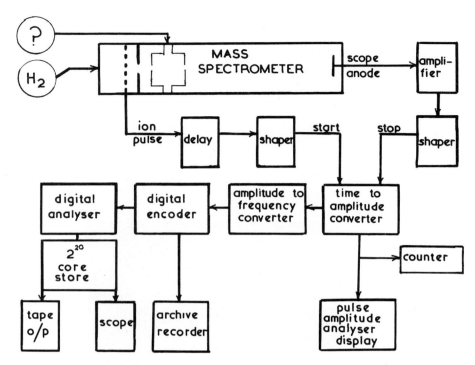

Fig. 2: Hydrogen ion spectrometer.

scanned. The output pulse passes on to a standard 100 channel pulse height analyser and is displayed in binary histogram form.[1] A choice of analyser sensitivity gives a range of channel widths from 0·2 nsec to 8 nsec. An event recorder measures the total ion count.

The input to the amplitude-to-frequency converter charges up a condenser which is then compared with a standard. The number of pulses from a pulse generator required to equalise these charges is a measure of the input amplitude, and the output is a frequency between 0 and 1 MHz. This passes through a gate and is counted and digitised by the encoder. The gate is an important resolution control. With the spectrometer ion pulse frequency of 10 kHz the maximum number of pulses admitted to the encoder would be 100, giving a resolution no better than the pulse height analyser. By holding this gate open for longer than 100 μsec and inhibiting the start pulse, any degree of increased resolution may be obtained. Resolution is taken here to be the time separation between adjacent information locations. The encoder has ten binary channels for flight time information together with five channels for labelling. Thus the theoretical maximum resolution for the whole system is with the 50 nsec ramp rise time displayed over 1023 data channels; about 0·05 nsec. The digitised flight times are kick sorted and placed in core store, each channel having a ten binary bit capacity. Core store can be dumped on tape and it is hoped to develop a continuous store scanning facility. The digitised data is also recorded for archive purposes.

The experimental system described above is currently complete except for the link between amplitude-to-frequency converter and the digitial encoder. Preliminary results quoted have been taken from the pulse amplitude analyser only.

Preliminary results

The H_2^+ – argon interaction is interesting in that there are only a limited number of reactions possible, all of which are well separated in energy change.

With the spectrometer pumped to 10^{-7} mm Hg, hydrogen was admitted to the source and the $(m/e = 2)$ peak monitored for 10^6 ions. For statistical reasons the probability of recording an ion event during one spectrometer cycle must be small (<1 in 10 cycles) so trap current and electron energy were kept to a minimum. The free ion velocity profile is derived from the pulse amplitude histogram, a typical peak being similar to Fig. 3. With argon at about 10^{-4} mm in the test section the run was repeated.

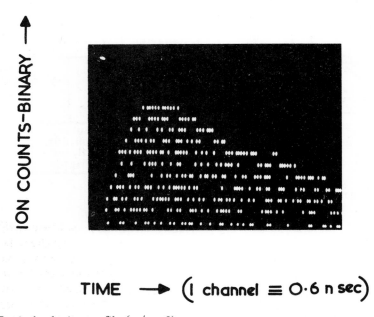

Fig. 3: Typical velocity profile $(m/e = 2)$.

After normalisation, the two histograms were subtracted (the proper deconvolution programmes not being ready) to yield the ion distribution shown in Fig. 4. To account for finite channel width this distribution was smoothed to give the scattered ion spectrum (Fig. 5), the precise positions of the peaks being found by polynomial fitting. Taking the position of the maximum of the free ion peak as an energy datum the energy shift corresponding to each peak can be readily determined.

To identify the processes associated with each peak in the ion spectrum it is necessary to consider all possible reactions between the H_2^+ ions and argon atoms. Due to the spectrometer geometry, only forward going and rearward going products of molecular disintegration will be detected. Thus any such reaction involving the source ion will give rise to two distinct energy losses. In this case the difference in translational energy between forward and rearward disintegration products of H_2^+ ions can be theoretically determined from the interaction potential curves of $H - H^+$ and $H^+ - H^+$.

The possible two-body collisions between H_2^+ ions and argon atoms with theoretical

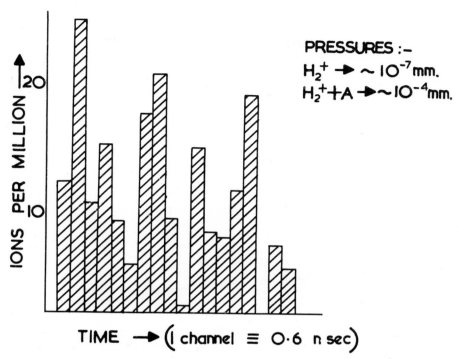

Fig. 4: Distribution of scattered ions H_2^+ on A.

kinetic energy changes of the ion[2] are:

(1)	$H_2^+ + A = H_2^+ + A^+ + e$	-15.7 eV	
(2)	$\quad\quad = H_2^+ + A^{2+} + 2e$	-27.6 eV	
		Forward	Backward
(3)	$= H^+ + H + A$	-5 eV	-10 eV
(4)	$= H^+ + H + A^+ + e$	-20.7 eV	-25.7 eV
(5)	$= H^+ + H + A^{2+} + 2e$	-32.6 eV	-37.6 eV
(6)	$= 2H^+ + e + A$	-23 eV	-30 eV
(7)	$= 2H^+ + 2e + A^+$	-38.7 eV	-45.7 eV
(8)	$= 2H^+ + 3e + A^{2+}$	-50.6 eV	-57.6 eV

Of the energy range being scanned in this experiment only the -5 eV process was not detected. Examination of ion densities in this region reveals a small change in slope, but the large number of unscattered ions swamp any small changes in intensity as a result of an inelastic scattering process. Success in the low energy shift region will require a less dispersed ion source or a more sophisticated mathematical analysis of the data.

The discrepancies between the measured energy shifts and those predicted theoretically need not concern us here other than to note that should both sets of figures be correct the difference must be attributable to transfer of kinetic energy. The ability to measure simultaneously both momentum and energy transfer could contribute to the understanding of energy sharing between internal and external modes.

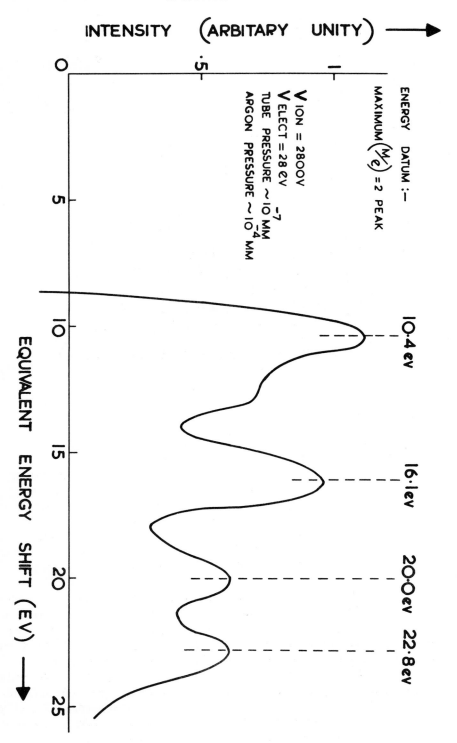

Fig. 5: Scattered ion spectrum H_2^+ on A.

Conclusion

In its final form this experimental system will have a theoretical energy resolution of about 0·01 eV and with an integration time limited only by the stability of the spectrometer or the associated electronics; highly infrequent processes (1 ion in 10^8) may be monitored for hours or longer. Instabilities in real time devices are critical, but with a fuller understanding of the error spectrum of the electron multiplier and the time-of-amplitude converter these could be included in a general deconvolution programme that could also account for ion source dispersion.

References

1   J. W. White, *Rev. Sci. Inst.* **38**, 2, 187 (1967).
2   H. S. Massey and E. H. Burhop, *Electronic and Ionic Impact Phenomena,* Clarendon Press, Oxford, 1952, p. 230.

Chapter 10

A Combined Field-Ion Microscope and Time-of-Flight Mass Spectrometer

P. J. Turner and M. J. Southon

Department of Metallurgy, University of Cambridge, England

Introduction

Müller and his co-workers have shown[1,2] that the combination of a conventional field-ion microscope with a time-of-flight mass spectrometer yields an instrument capable of revealing both the atomic structure of a solid surface and the chemical identity of individual atoms observed in the field-ion image. The present article describes the design and performance of a similar instrument.

In the conventional field-ion microscope[3-6] the specimen is in the form of a sharp point of a few hundred ångstroms radius of curvature. When a positive potential of several kilovolts is applied to the specimen in the presence of a few millitorr of an inert gas, the intense electric field of a few volts per ångstrom generated at the specimen point is sufficient to cause field-ionisation of the gas by electron tunnelling from gas atoms into the adjacent specimen surface. Field-ionisation has been employed in ion-sources for mass spectrometric analysis of gases or investigation of surface reactions,[7,8] but for field-ion microscopy conditions are such that ionisation of the 'imaging gas' only occurs over individual atoms protruding slightly from the curved surface of the specimen. The ion beam diverging from the specimen is allowed to strike a fluorescent screen where the resulting pattern constitutes an image of the specimen surface. Figure 1 is a typical image of a tungsten specimen in which each bright spot corresponds to an individual tungsten atom at the surface of the specimen, each ring of spots corresponds to the edge of a single plane of tungsten atoms, and each family of rings represents a particular crystallographic pole on the roughly hemispherical specimen surface.

With this ability to resolve atomic structure directly, field-ion microscopy has been widely used in investigations of surface phenomena and microstructure in metals and alloys[3-6]. The information contained in the field-ion image is essentially structural, giving little information on the nature of the atoms which give rise to individual image points. For example, Fig. 2 is a micrograph of a specimen of an alloy steel, obtained using neon as the imaging gas. The surface is less regular than the tungsten surface of Fig. 1, with a number of particularly bright image points which may correspond to impurity atoms, misplaced iron atoms or other point defects: the bright band across the bottom of the micrograph is the edge of a platelet of vanadium carbide precipitate. The chemical identity or composition of such features, as well as their structure, is of particular interest to

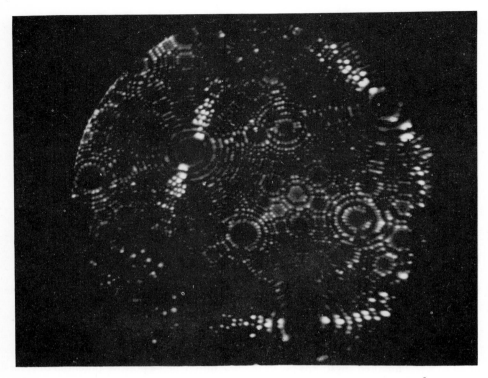

Fig. 1: **Helium**-ion micrograph of a tungsten specimen of approximately 350 Å radius of curvature.

metallurgists and this information may be ascertained, as Müller and others have shown,[1,2,9] by mass-analysis of the ions field-evaporated from the specimen surface.

If the potential applied to the specimen is increased above the level required for image formation, a well-defined critical value can be reached at which surface atoms of the specimen itself are removed as positive ions. This process, known as field-evaporation,[10] is related to field-ionisation but occurs by thermal evaporation of a surface ion over a potential barrier reduced almost to zero by the applied field. From the simple representation of Fig. 3 it can be seen that the activation energy Q for evaporation of an atom as an ion of charge ne from a surface subject to a field strength F is given approximately by

$$Q = Q_0 - (n^3 e^3 F)^{1/2}; \qquad (1)$$

Q_0, the binding energy of the ion in the absence of a field, is given by

$$Q_0 = \Lambda + \sum_n I_n - n\phi_0 \qquad (2)$$

where ϕ is the sublimation energy of the evaporating atom, $\sum_n I_n$ is the sum of the first n ionisation energies of the atom and m_0 is the average work function of the surface. The rate of field evaporation k_e may be represented by the usual Arrhenius equation

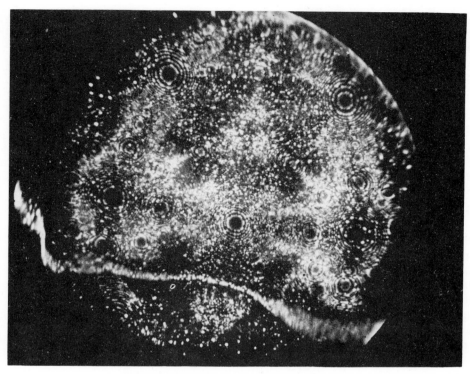

Fig. 2: Neon-ion image of a specimen of an alloy steel. The bright band across the bottom of the micrograph corresponds to the intersection of a platelet of vanadium carbide precipitate with the specimen surface.

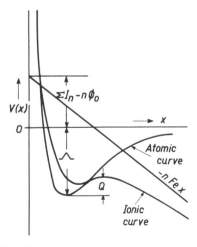

Fig. 3: Potential diagram for field-evaporation. The atomic curve represents the potential energy $V(x)$ of a neutral atom as a function of distance x from the specimen surface. The corresponding curve for an ion is depressed by the applied field F to leave only a small potential barrier of height Q, described by Eqns. (1) and (2). (After Brandon[21].)

$$k_e = \nu \exp(-Q/kT) \tag{3}$$

where ν is the frequency of vibration of a surface atom, and the dependence of the rate of evaporation on the field strength F is described by[11]

$$\frac{\partial \ln k_e}{\partial \ln F} \sim \frac{1}{2}\frac{Q_0}{kT} \tag{4}$$

Since $Q_0 >> kT$ for the refrigerated specimen in the field-ion microscope, it is clear that the rate of field-evaporation is sensitively dependent on the field strength. Experimental values of the order of 300 have been measured[12,13] for the derivative defined in Eqn. (4), implying that an increase in field strength of only 7% will increase the rate of field-evaporation by a factor of 10^9.

It is, therefore, feasible for a fraction of an atomic layer to be field-evaporated from the specimen surface during a high-voltage pulse of the order of nanoseconds duration, rather than in the more usual time of several seconds. Measurement of the time taken by the evaporated ions, of known energy, to traverse a path of known length to a suitable detector then clearly yields the value of the mass-to-charge ratio of the ions; this is the principle of the time-of-flight mass spectrometer. The area on the specimen surface from which ions are sampled is defined by recording only those ions which pass through a small aperture in the fluorescent screen which displays the field-ion image: Müller[14] and Brenner and McKinney[15] have shown that the evaporated surface atoms follow almost the same trajectories as the gas ions which formed their images prior to evaporation. This sample area can be as small as $100Å^2$ or less, only a few atomic areas, and the particular advantage of time-of-flight spectrometry is that all possible mass numbers may be monitored simultaneously for each individual ion evaporated from this small area. As Müller has pointed out,[16] since the detector signal need only be recorded over the limited range of possible arrival times of the evaporated ions, the severe noise problem inherent in other methods[17] of analysing small numbers of field-evaporated ions is avoided.

Experimental

The apparatus which has been built for the present work is shown schematically in Fig. 4. The specimen is mounted at the end of a cold finger which may be manipulated to a small extent by means of a bellows and gimbals in order to bring any part of a small region of the field-ion image over the defining aperture, of 4 mm diameter, in the fluorescent screen. The microscope section of the apparatus includes an ion-to-electron image converter system:[18] the field-ion image is converted to a secondary-electron image at a fine mesh, with the advantage of avoiding the poor efficiency of the phosphor for heavier ions and thereby greatly facilitating microscopy of less-refractory specimen materials, as described elsewhere.[19] The final image is viewed and photographed through a vacuum window via a 45° front-silvered mirror. Ions passing through the aperture in the fluorescent screen and the further aperture in the mirror traverse a flight-tube to strike the detector, which is a Bendix M306 magnetic electron multiplier. In addition to a short time-constant and stability to atmospheric exposure, this type of multiplier has the particular advantage of a flat cathode of $2 cm^2$ area: the uncertainty in the ion path length arising with detectors with inclined electrodes, such as the venetian blind or tubular continuous-

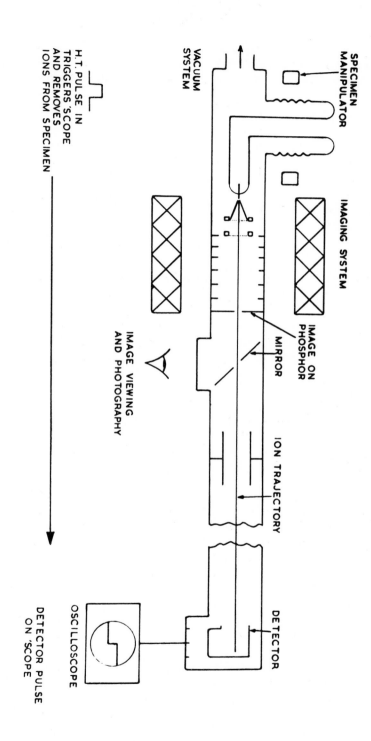

Fig. 4: Schematic diagram of the combined field-ion microscope and time-of-flight mass spectrometer.

dynode types, is thereby eliminated. Although the detection efficiency of the multiplier is probably near 100%, the presence of two meshes in the ion path limits the transmission of the system to about 40%.

The short field-evaporation pulse is provided by a Huggins 961EA pulse-generator. Figure 5 shows the shape of the pulse; the pulse amplitude is continuously variable from

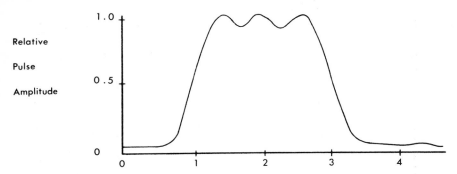

Fig. 5: Shape of the field-evaporation pulse: relative pulse-height against time in nanoseconds.

zero to 6 kV, the rise and fall times are less than 0·5 nsec and the minimum pulse-width is 2 nsec. The fact that the pulse is not flat-topped is not serious since, with the very large voltage-sensitivity of the field-evaporation rate, evaporation is likely to occur only at the peak pulse voltage. Capacitative coupling is used to superimpose the pulse upon the d.c. high voltage applied to the specimen. Although the pulse must be reflected to some extent in the high-voltage line, which is not terminated at the specimen, there is no experimental evidence that reflected pulses are of sufficient amplitude at the specimen to cause further field-evaporation. The reproducibility of the pulse-height is claimed by the manufacturers to be better than 4%, and the stability of the d.c. high-voltage supply, a Brandenburg S0530/10, is of the order of 0·1%.

A Hewlett-Packard 180 oscilloscope is used for measurement of flight times. The single sweep of the oscilloscope is triggered synchronously with the pulse by the radiative signal from the pulse-generator, and the output pulse from the electron multiplier is fed directly to the y-plates of the oscilloscope, which has a rise-time of 7 nsec. The line from the multiplier to the oscilloscope is not terminated at the oscilloscope: the length of this line is therefore minimised at 5 cm to ensure that the time taken for attenuation of reflected pulses is short compared with the rise-time of the oscilloscope. The multiplier pulse then appears as a step on the oscilloscope trace since the decay time of the circuit is long, being of the order of 40 μsec. The trace is recorded photographically using an f1·8 lens and a fast green-sensitive emulsion, Agfa-Gevaert Scopix G. It was found that when the line was terminated at the oscilloscope the trace of the multiplier pulse could not be detected photographically.

The apparatus is evacuated by two liquid-nitrogen cooled, oil-diffusion pumped systems and is not bakeable. The ultimate pressure in the microscope chamber is typically 5×10^{-8} torr. The presence of imaging gas at the usual pressure of 3×10^{-4} torr would give the possibility of ion feedback in the electron multiplier, resulting in spurious output pulses. The system is therefore arranged to provide some differential pumping for the flight-tube and the imaging gas pressure there is kept below 5 x 10^{-5} torr; with the multiplier gain at 10^6 no ion feedback is observed. In most instances mass-analyses are carried out in the absence of the imaging gas, since the partial pressure of active gases introduced with the imaging gas might exceed the background pressure of 5 x 10^{-8} torr.

The distance between the specimen and the cathode of the electron multiplier is 1·61 m. The flight-time t of an ion of mass m a.m.u. and charge ne may then be written as

$$t = 3\cdot73 \, (m/nv)^{1/2} \text{ microseconds} \tag{5}$$

where v kilovolts is the total potential applied to the specimen. Typical flight times lie in the range from 1 to 20 μsec with voltages of up to +15 kV applied to the specimen. Flight times less than 0·5 μsec are difficult to measure because of disturbance of the oscilloscope by the radiative signal from the high-voltage pulse. The phosphor screen and flight tube are grounded and the meshes in the image converter are operated at −8 and −7 kV: a small correction, shown in Fig. 6, must therefore be made to the flight-time described by Eqn. (5).

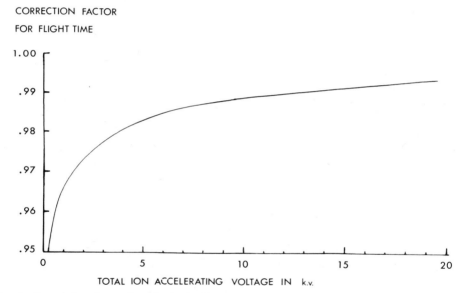

Fig. 6: Correction factor by which the flight-time described by Eqn. (5) must be multiplied to allow for acceleration of the ions through the image-converter section of the apparatus.

With the oscilloscope sweep rate at the maximum usable value of 5 cm/μsec, measurement of photographic records of traces gives flight times to an accuracy of better than 10 nsec. However, such accuracy is only meaningful if the ion energy is sufficiently well-defined. Differentiation of Eqn. (5) gives

$$\frac{\delta m}{m} = \frac{2\delta t}{t} + \frac{\delta v}{v} \tag{6}$$

Under typical conditions with a d.c. voltage of 10 kV applied to the specimen and a pulse of 1·5 kV, $v = 11·5$ kV and $t = 7·8\,\mu\text{sec}$ for an ion of $m/ne = 50$, $2\frac{\delta t}{t}$ is then of the order of 1/400, whereas if δv is determined by the 4% jitter in the pulse height, $\frac{\delta v}{v}$ is of the order of 1/200. It therefore appears that jitter of the pulse height is at present limiting the mass resolution. The effect can be minimised by the use of small pulses, and in practice resolutions $\frac{\delta m}{m}$ of better than 1/300 have been achieved, suggesting $\delta t \sim 10$ nsec and $\delta v \sim 20$ volts. On a single trace, the time difference between the arrival of two ions of different m/ne may be measured to better than 10 nsec, but the accuracy of interpretation is limited by the fact that the transit time for electrons across the cathode of the multiplier is about 5 nsec. When comparing different traces the jitter in the triggering of the oscilloscope should be considered: this is claimed by the manufacturers to amount to only 2½ nsec in the delayed mode of operation generally used in this work, and may be neglected.

Results and discussion

Figure 7 is a field-ion image of a tungsten specimen obtained in the apparatus des-

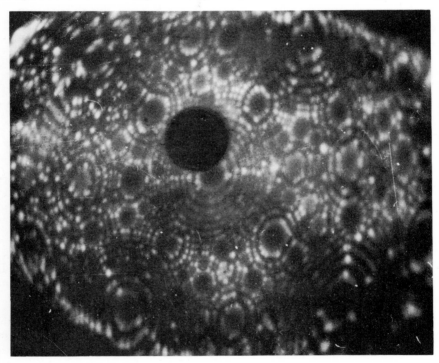

Fig. 7: Helium-ion image of a tungsten specimen recorded in the present apparatus, showing the defining aperture magnified with respect to the image by a linear factor of 3.

cribed above. The dark disc near the central (110) pole corresponds to the aperture in the fluorescent screen but, since the screen is not in the plane where the ion image is initially formed in the image converter system (Fig. 4), the diameter of the disc must be demagnified by a factor of 3 to obtain its size relative to the ion image. In the case of Fig. 7 the effective aperture diameter referred to the specimen is about 20 Å.

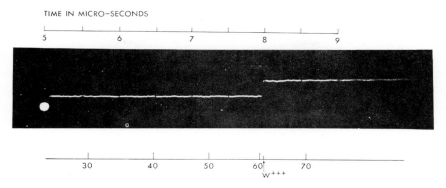

Fig. 8: Oscilloscope trace showing the result of a single field-evaporation pulse applied to a tungsten specimen. The trace is initiated by the field-evaporation pulse with a delay of 5 μsec; the step on the trace is due to the arrival of a single W^{3+} ion, $m/ne = 60.5$, after a flight-time of 8.0 μsec. The sweep rate is 2 cm/μsec: the centimetre graduations on the oscilloscope screen can be seen on the trace.

Figure 8 shows an oscilloscope trace resulting from the application of a single field-evaporation pulse to a tungsten specimen with one W^{3+} ion arriving at the detector. The measured mass-to-charge ratio is 60.5, suggesting ^{182}W. An appreciable proportion of W^{4+} ions have also been observed; for field-evaporation near the centre of the (110) pole the ratio of W^{3+} to W^{4+} was of the order of 3. This is a surprising result since Brandon[20] has predicted that tungsten should evaporate as W^{2+}, and also since the simultaneous occurrance of two relatively high charge-states, also observed by Müller and co workers[2] and by

Table 1. The charge states of the metal ions field-evaporated from the materials which have been studied so far in the present work.

Material	Atomic Weights	Charge Observed	Charge Calculated[20]
W	(180), 182, 183, 184, 186	3+, 4+	2+
Mo	92, 94, 95, 96, 97, 98, 100	2+, 3+, 4+	2+
Ir	191, 193	2+, 3+	2+
Co	59	2+	2+
Pt	(190), (192), 194, 195, 196, 198	2+, 3+	2+
Ti	46, 47, 48, 49, 50	2+	2+
Ni	58, 60, (61), (62), (64)	2+	2+
Fe	54, 56, (57), (58)	2+	2+
Au	197	1+, 2+, 3+	2+
Cu	63, 65	1+, 2+	1+

Brenner and McKinney,[9] cannot be explained[9] by the present theories. A relatively large proportion, of the order of 50%, of the ions field-evaporated from tungsten are found to be molecular ions which may be oxides, nitrides or carbides but which are difficult to identify unambiguously because tungsten itself has four isotopes of comparable abundance. Although Müller has observed[2] similar ions under relatively modest vacuum conditions of 10^{-6} torr, the occurrence of such a large proportion of molecular ions at a pressure of 5×10^{-8} torr must cast doubts upon the usual claim of field-ion microscopists (e.g. reference 19) that specimen surfaces are clean and remain clean indefinitely under the very high field strengths used for helium-ion microscopy. Similar proportions of molecular ions have also been observed during field-evaporation of molybdenum and iridium specimens.

Results for cobalt and platinum were obtained with an equiatomic Co-Pt alloy specimen, for cobalt and titanium with a Co-10%Ti alloy, and the remainder with the pure metal. The atomic weights in parentheses refer to isotopes whose natural abundance is less than 5%.

It is seen that most metals evaporate as the predicted doubly charged ion, with W, Mo, Ir and Pt occurring in higher charge states. Gold also gives a triply charged ion, but

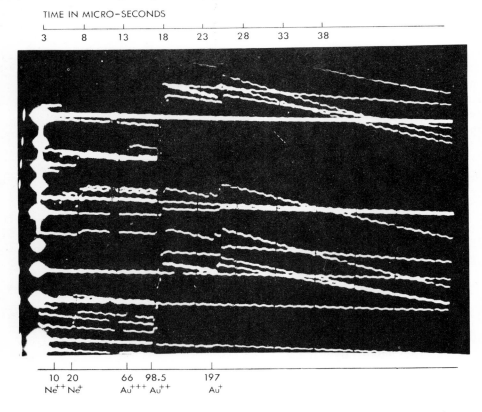

Fig. 9: Field-evaporation products from a gold specimen, each trace representing the result of a single evaporation pulse. The large signals at $m/ne = 98.5$ indicate the production of many Au^{2+} ions per pulse because of field-enhancement following deformation of the specimen. Adsorbed neon is also removed by field-evaporation. The ripple on the traces is noise from an external source.

only under exceptional conditions. The large mechanical stress associated with the strong electric field required for field-evaporation not infrequently causes gold specimens to undergo plastic deformation by slip. The specimen surface then becomes stepped and the field strength at a given applied voltage is enhanced at the steps. When a further high-voltage pulse is then applied, the field strength at the steps can rise well above the critical value required to field-evaporate a few atoms in the 2 nsec pulse duration. Figure 9 shows an example of this, where it can be seen that several traces show very large signals (and correspondingly steep decays) at $m/ne \sim 9.5$, signifying the evaporation of many Au^{2+} ions during a single pulse. The pulse-height distribution from the multiplier for single ions is so wide that the number of ions producing a particular pulse height cannot be estimated, but it was evident that Au^{3+} ions only occurred on traces which showed very large Au^{2+} signals. It may be inferred that Au^{3+} only occurs at exceptionally high field strengths and large evaporation rates, which suggests in turn that gold may initially field-evaporate as Au^+ or Au^{2+} with a transition to Au^{3+} by field-ionisation as the particle leaves the surface if the field is sufficiently high. Muller has already speculated[16] as to whether this mechanism is generally responsible for the surprisingly high charge states produced by this technique.

It may also be noted in Fig. 9 that the time spread of the Au^{2+} pulses is relatively large, of the order of a microsecond. This time-spread corresponds to a voltage-spread of

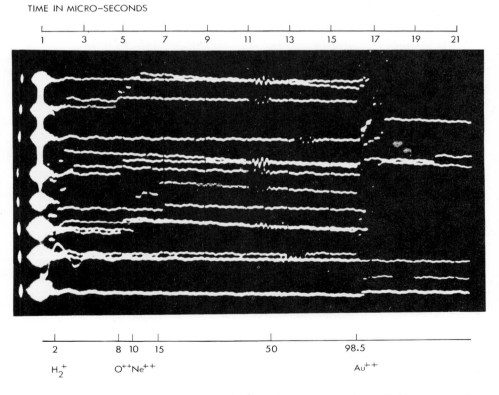

Fig. 10: Faster traces show the presence of H_2^+ in the spectrum of ions field-evaporated from gold. Hydrogen is only observed after plastic deformation of the specimen.

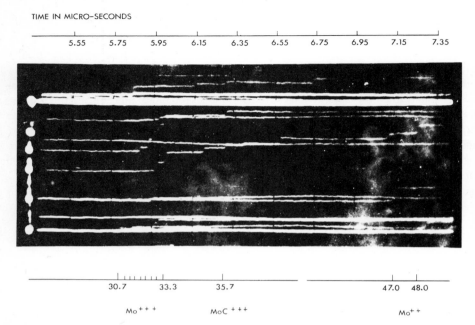

Fig. 11: Field-evaporation products from molybdenum recorded at an oscilloscope sweep rate of 5 cm/μsec. Within the range of Mo^{3+} isotopes, separations of 30 nsec between pulses demonstrate the resolution of adjacent mass numbers.

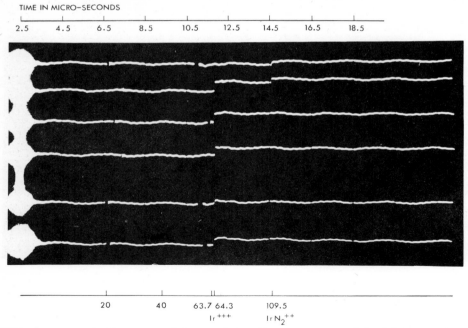

Fig. 12: Some ions field-evaporated from iridium. Close inspection of the 5 Ir^{3+} ions shows that 2 are ^{191}Ir and 3 are ^{193}Ir, with a separation of the order of 60 nsec between the isotopes.

0·6 kV: comparison with the pulse height of 0·7 kV shows that field-evaporation occurred at almost the whole range of voltages covered during the pulse.

Figure 9 also shows the presence of Ne^+ and Ne^{2+} in the mass spectrum, and Ne^{2+} is also evident in Fig. 10, indicating that neon is absorbed on gold at the specimen temperature, $\sim 78°K$, since although neon was used as the imaging gas for gold it was removed several minutes before the data of Figs. 9 and 10 were obtained. Figure 10 also shows the presence of H_2^+ only on traces which also show a very large Au^{2+} signal. H_2^+ has also been found after plastic deformation has occurred during field-evaporation of copper, nickel and iridium, but the origin of the hydrogen is not clear.

Figure 11 shows results obtained with a molybdenum specimen using a sweep speed of 5 cm/μsec in order to demonstrate the resolution of the instrument. The group of ions arriving at the lower end of the time-scale are triply-charged molybdenum isotopes, with Mo^{2+} at the higher end of the scale. Mo^{4+} has occasionally been observed when large pulse-heights have been used. The measured ratio of occurrence is approximately Mo^{4+} : Mo^{3+} : Mo^{2+} 1 : 3 : 5. Within the Mo^{3+} group, measurement of the various time intervals between the arrivals of pairs of ions shows that the intervals are integral multiples of 30 nsec, which is the interval expected between adjacent mass numbers for Mo^{3+}. A few ions arrive at times which do not fit this scheme: a possible explanation is that the energy of

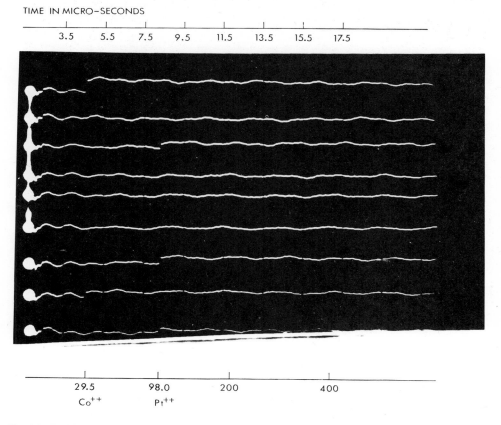

Fig. 13: Field-evaporation of an alloy, equiatomic Co-Pt.

these ions is slightly lower than the energy of field-evaporated Mo^{3+} because they have been formed by field-ionisation of Mo^{2+} at a small distance from the specimen surface. The expected energy deficit for this process is about 15 eV, corresponding to a time-lag of some 12 nsec.

Some of the species field-evaporated from iridium are shown in Fig. 12. Both Ir^{2+} and Ir^{3+} are found, together with IrN_2^{2+}, IrO_2^{2+} and IrO^{2+}. The occurrence of these species is completely reproducible. Figure 13 shows some of the results obtained during field-evaporation of an equiatomic cobalt-platinum alloy, from which the species observed are Co^{2+}, Pt^{2+} and Pt^{3+}. These results were obtained during an analysis of a sample of area 50 Å² and depth 25 Å : the results were consistent with the assumption that each multiplier pulse corresponded to the arrival of a single ion.

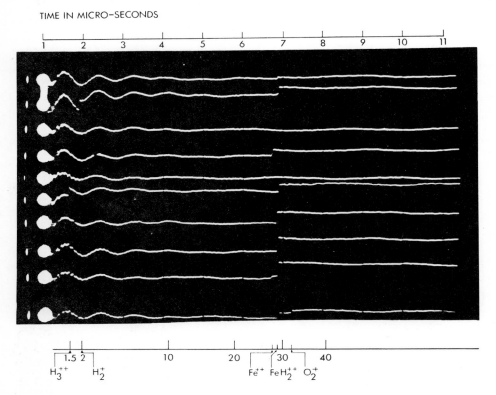

Fig.14: Field-evaporation of a specimen of pure iron. Field-evaporation was carried out at a pressure of 5×10^{-8} torr, but hydrogen at a pressure of 1×10^{-4} torr had previously been used to etch the specimen surface before imaging.

Figure 14 shows some results obtained with a specimen of high-purity iron, as a preliminary to a current experiment intended to investigate segregation of phosphorus to grain boundaries in iron. The figure is included here to demonstrate the applicability of this technique to materials of major metallurgical and technological importance.

Conclusion

The instrument described in this article is both a versatile field-ion microscope, particularly well-suited to imaging less refractory specimen materials, and a time-of-flight mass spectrometer, giving a mass analysis of individual atoms observed in the field-ion image. It has been shown that the mass resolution of the instrument is sufficient for separation of adjacent isotopic masses for the ions that have been encountered. This type of instrument clearly has a wide range of applications in fine-scale microscopy and microanalysis, but a prerequisite to its reliable use is a knowledge of the types of ions to be expected under the unusual experimental conditions. The preliminary results reported in this article are a contribution to this knowledge.

Acknowledgements

This work has been supported in part by the Science Research Council and by the UK Atomic Energy Authority. One of the authors (P.J.T.) is grateful to SRC for personal support.

References

1. E. W. Müller and J. A. Panitz, *14th Field Emission Symposium, Gaithersburg, Maryland, 1967.*
2. E. W. Müller, J. A. Panitz and S. B. McLane, *Rev. Sci. Instr.* **39**, 83 (1968).
3. E. W. Müller, *Advan. Electron. Electron Phys.* **13**, 83 (1960).
4. *Field-Ion Microscopy* (Eds. J. J. Hren and S. Ranganathan) (proceedings of a short course held at the University of Florida, 1966); Plenum Press, New York, 1968.
5. *Applications of Field-Ion Microscopy* (Eds. R. F. Hochman, E. W. Muller and B. Ralph) (proceedings of a Conference held at Georgia Institute of Technology, 1968); to be published.
6. E. W. Müller and T. T. Tsong, *Field-Ion Microscopy,* Elsevier, New York, 1969.
7. H. D. Beckey, *Advan. Mass. Spectr.* **2**, 1 (1963).
8. H. D. Beckey, H. Knoppel, G. Metzinger and P. Schulze, *Advan. Mass Spectr.* **3**, 35 (1966).
9. S. S. Brenner and J. T. McKinney, *Appl. Phys. Letters* **13**, 29 (1968).
10. E. W. Müller, *Phys. Rev.* **102**, 618 (1956).
11. D. G. Brandon, *Surface Sci.* **3**, 1 (1964).
12. D. G. Brandon, *Brit. J. Appl. Phys.* **16**, 683 (1965).
13. D. M. Taylor, private communication.
14. E. W. Müller, *15th Field Emission Symposium, Bonn, 1968.*
15. S. S. Brenner, and J. T. McKinney, *15th Field Emission Symposium, Bonn, 1968.*
16. E. W. Müller, *Chem. Soc. Quart. Rev.* **23**, 177 (1969).
17. D. F. Barofsky and E. W. Muller, *Surface Sci.* **10**, 177 (1968); *Intern. J. Mass Spectrom. Ion Phys.* **2**, 125 (1969).
18. P. J. Turner and M. J. Southon, *15th Field Emission Symposium*, Bonn, 1968.
19. P. J. Turner and M. J. Southon, *Proc. 4th Int. Vacuum Congr., Manchester, 1968;* Institute of Physics, London, p. 207, 1969.
20. D. G. Brandon, *Phil. Mag.* **14**, 803 (1966).
21. D. G. Brandon, *Advan. Optical Electron Microscopy* **2**, 343 (1968).

Discussion

R. I. Reed: After almost every slide you call attention to 'events' attributable to surface impurities. M. Inghram made an extensive use of arc-imaging to clean surfaces in his investigations; have you considered the use of this technique to clean the surface before you make your own experiments?

P. J. Turner: It is difficult to use thermal desorption of impurities to clean specimens for field-ion microscopy, since surface diffusion at elevated temperatures tends to blunt the specimen to a prohibitive extent. It is probable that field-evaporation cleans the specimen apex in most cases, but it is not clear to what extent the apex becomes contaminated by impurities diffusing from the shank of the specimen.

Chapter 11

The Use of Time-of-Flight Mass Spectrometry as a Selective Detector for Quantitative Gas Chromatography

C. D'Oyly-Watkins, D. E. Hillman, D. E. Winsor and R. E. Ardrey

Chemical Inspectorate, Woolwich, England

Introduction

The advantages of selective gas chromatographic detectors are well known in the pesticide field. The use of the electron capture detector enables minute traces of active material to be estimated in the presence of large amounts of solvent, the peaks from which would normally completely obscure that of the pesticide.

Unfortunately there are few such selective detectors. The only two in general use are the electron capture one for halogenated and some oxygenated compounds and the thermionic detector for phosphorus compounds. The advantages of a detector having variable selectivity which could be made to respond only to one class of compounds or even a single compound in a mixture are obvious. Apart from the identification and estimation of minor components in complex mixtures as described above, many other estimations are difficult due to partial or total lack of separation on the column in use. Although a column can usually be found which will produce the desired separation, the exercise is often very time consuming and can be avoided by the use of a selective detector.

Mass spectrometry has for some time now been used in conjunction with gas chromatography for the identification of the separated components. In the early days of the combined technique, Henneburg and Schomburg even suggested the use of a mass spectrometer monitoring continuously a single mass ion abundance. The instrument gave a chromatogram showing peaks for only those compounds which produced the selected ion on fragmentation. By this means, information could be obtained from spectrometers which did not have sufficient scan speed to produce a spectrum during the short emergence time of a chromatographic fraction.

A number of workers have since devised means of monitoring a fraction of the total ion current in mass spectrometers so that they act as general gas chromatograph detectors. This removes the need for a separate detector on the chromatograph and eliminates any time-lag between the two instruments when used for qualitative identification of separated components. Satisfactory quantitative estimations of completely separated components have been achieved by this means.

The main difficulty in the use of the mass spectrometer as a selective detector lies in the impossibility of accurately repeating the size of the minute sample injected on successive runs. In order to obtain a reasonable degree of precision, all measurements required must be made on a single chromatographic run. For the estimation of two unseparated components in a chromatographic fraction, it is necessary to monitor two ion abundances. A third could be required if an internal standard were used, to enable absolute values rather than ratios to be calculated for the two components.

Ion abundance curves for this purpose were obtained by Lindeman and Annis[2] who plotted intensities of the required m/e values against time, using data from multiple mass spectra, recorded every few seconds during the elution of the fraction. Another, more automated, technique has since been described by Sweeley et al.[3] They employed a time-actuated relay and voltage divider to switch the accelerating voltage of a conventional mass spectrometer rapidly between two levels to produce a composite recording of the changes in intensities of two preselected m/e values. This device is now commercially available in a form which permits monitoring of three ion abundances simultaneously. The amount by which the accelerating voltage can be varied is unfortunately restricted, so that the m/e values employed can be only a few mass numbers apart. This, while being quite satisfactory for the determination of stable isotopic abundances in eluted compounds, limits the value of the system for general application as a selective detector.

The time-of-flight mass spectrometer, with its ability to monitor simultaneously six ion abundances anywhere in the spectrum, is obviously well suited for this application. Some preliminary work by the authors[4] on its use for this purpose showed sufficient promise to justify the more detailed evaluation of its performance.

The gas chromatograph-mass spectrometer interface

The interconnecting system between the two instruments is shown in Fig. 1. The gas chromatograph used was a Pye Panchromatograph and the mass spectrometer was a Bendix 12-107 fitted with five Type 3012 analogue scanner units. The carrier gas stream

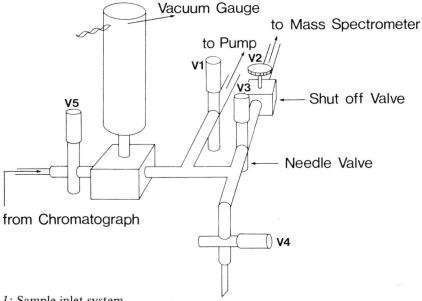

Fig. 1: Sample inlet system

from the chromatograph column was split so that part went to the normal flame ionisation detector and the rest through a heated 0·5 mm stainless steel capillary and needle valve (V.5.) into the mass spectrometer inlet system. In the design of this system, the use of sample concentration devices such as those of Ryhage and Watson and Bieman have been avoided since they depend for their operation on mass discrimination effects. The split produced by such methods is, to some extent, effected by operating parameters such as sample concentration. Their inclusion would therefore reduce quantitative accuracy.

The sample was admitted to the mass spectrometer through a calibrated needle valve (V.3.). A shut-off valve (V.2.) was included in the line so that the mass spectrometer could be isolated without affecting the setting of the needle valve. Reduction of the pressure in the sample system by means of an auxilliary pumping system was found to be necessary, since restriction of the gas flow to a level at which the spectrometer vacuum could be maintained, by shutting down V.5., increased the amount of sample passing through the flame ionisation detector at the expense of that reaching the spectrometer. Furthermore, an excessive amount of peak tailing was found to occur due to the slow rate at which the concentration built up in the sampling system was reduced by pumping with the spectrometer alone. A Pirani gauge for measurement of the intermediate vacuum was included in the system.

The side-arm, containing V.4. and terminating in an hypodermic needle, was added so that the spectrometer could be used separately for the examination of volatile samples. It has also proved useful for directly admitting a sample of a mixture to be analysed so that the analogue scanner gates could be set on the appropriate mass peaks. This operation is difficult to perform with the constantly changing concentration of sample in the chromatograph effluent.

At the time when these measurements were made, the sample system was constructed with ¼″ copper tubing, Swagelok connections and Nupro valves. It was heated by tape wrapped round the piping. It has since been redesigned, using stainless steel tubing and all-metal bellows valves, to eliminate 0-rings from the system. The Panchromatograph has been replaced by a Perkin-Elmer F.11 chromatograph which is small enough to be mounted close to the sample inlet system. This enables the long capillary connecting tube to be replaced by a short length of ¼″ stainless steel tubing. The whole system has been enclosed in a thermostat-controlled oven, mounted adjacent to the chromatograph oven and extending as near as possible to the spectrometer source. The Pirani gauge has been omitted and the various valves mounted close to each other to minimise the length of connecting tubing. The reduction of detector volume, elimination of cold spots and increase of the temperature at which the system can be maintained, have significantly increased the stability, chromatographic resolution and the boiling point of samples which can be analysed.

Accuracy and reproducibility

In order to assess the performance of the mass spectrometer as a detector, four mixtures (A to D) of benzene, toluene and ethylbenzene in known amounts were prepared. The concentration of ethylbenzene was kept constant, while those of benzene and toluene were varied. Two analogue scanner gates were set so that intensities of m/e 78 and 91 could be recorded. The former was used for estimation of benzene and the latter for estimation of toluene and ethylbenzene. Six 0·25 µl samples of each solution were injected successively onto a 3 ft by 4 mm column containing 10% Apiezon L. The column temperature was 125°C. Well-resolved peaks of good shape were obtained, both on the chromatograph recorder and on the u.v. recorder connected to the mass spectrometer scanners. The heights of these were measured. Peak height rather than peak area measurements were used throughout, since they can be made easily and accurately. Peak height measurements have been found to give good results using the flame ionisation detector with the materials employed. The two detectors being run in parallel enabled results from individual runs to be compared directly; errors due to faults in sample injection, selective evaporation of sample components being common to both. The compositions of the mixtures, interface parameters and the results obtained by calculating ratios of peak heights for each run are shown in Table 1.

Table 1[a].

	A (%)	B (%)	C (%)	D (%)
Benzene	33·32	33·58	22·22	11·08
Toluene	33·36	35·11	44·44	55·58
Ethylbenzene	33·32	33·31	33·34	33·34

Solution	Detector[b]	Benzene/Ethylbenzene		Toluene/Ethylbenzene		Benzene/Toluene	
		Mean[c]	Variance	Mean[c]	Variance	Mean[c]	Variance
A	M.S.	1·619	15·04	1·018	3·83	1·587	12·43
	F.I.	0·626	2·60	0·416	1·03	1·405	2·28
B	M.S.	1·560	12·44	1·047	3·17	1·493	13·73
	F.I.	0·619	2·75	0·473	1·25	1·309	2·21
C	M.S.	1·151	9·92	1·361	2·59	0·815	9·91
	F.I.	0·483	0·90	0·533	1·56	0·874	3·90
D	M.S.	0·583	13·27	1·656	3·89	0·355	11·21
	F.I.	0·269	1·45	0·637	1·81	0·423	1·18

[a] Masses, benzene 78 toluene and ethylbenzene 91
Sample size: 0·25 litre
Intermediate pressure: 0·4 torr
Inlet leak; 50
[b] M.S.: Mass spectrometer; F.I.: Flame ionisation
[c] Means are of 6 results for each solution

These results were disappointing as they showed a very wide spread and gave coefficients of mass spectrometer determination variation of up to 15·04 for a run on which the flame ionisation detector gave a value of only 2·60. Further examination of the results showed that in the case of the toluene/ethylbenzene ratio, both peaks of which were recorded from a single analogue scanner, the spread was much smaller. Examination of the scanner gate pulses on the oscilloscope showed that the one monitoring m/e 78 was fluctuating slightly. Replacement of a faulty valve appeared to stabilise the gate. On re-running the same series of samples, using the scanner gates set as wide as possible, ratios were obtained for the benzene/ethylbenzene and benzene/toluene mixtures, giving spreads comparable with those for toluene/ethylbenzene. The coefficients of variation were nevertheless some three times greater than those obtained from the flame ionisation detector figures. In an attempt to improve this situation, further runs were made using mixture A and the spectrometer detector only, to see: (a) if there was an optimum sample size, (b) if a particular setting of the interface parameters would produce more stable operation of the system. Table 2 shows the effect produced by using three different sample sizes. It is clear that although an optimum value was found, it agreed with that of 0·25 µl originally chosen.

Table 2. Effect of sample size variation on spread of results[a].

Sample Size	Benzene/Ethylbenzene Mean[b]	Variance	Toluene/Ethylbenzene Mean[b]	Variance	Benzene/Toluene Mean[b]	Variance
0·20	1·489	5·78	1·010	3·91	1·474	3·41
0·25	1·023	3·06	1·013	2·78	1·012	2·42
0·30	1·427	6·10	0·986	3·80	1·499	6·87

[a] Inlet leak: 50
   Intermediate pressure; 0·4 torr
[b] Means of 10 results

Table 3. Effect of inlet leak variation on spread of results[a].

Inlet leak	Benzene/Ethylbenzene Mean[b]	Variance	Toluene/Ethylbenzene Mean[b]	Variance	Benzene/Toluene Mean[b]	Variance
25	1·157	2·81	0·980	3·40	1·182	4·17
50	1·137	2·50	1·061	2·04	1·091	1·18
75	1·159	3·44	1·082	0·86	1·071	2·59
100	1·127	9·45	1·083	4·62	1·039	5·35

[a] Intermediate pressure: 0·4 torr
[b] Means of 3 results each

The effect of varying the spectrometer inlet leak by adjustment of valve V.3. was examined. This gave the results shown in Table 3. The inlet leak figures quoted are the readings of the vernier scale of the valve.

It can be seen that larger leak rates produce a greater spread in results. This is probably due either to instability in the sample system or to other effects such as multiplier saturation in the mass spectrometer. Again the value of 50 originally chosen appears to be satisfactory.

Valve V.1. was closed slightly, so that the intermediate pressure increased to a value of 0·5 torr and mixture A was again run, giving the result shown in Table 4.

Table 4. Effect of intermediate pressure variation on spread of results.

Intermediate pressure (torr)	Benzene/Ethylbenzene Mean[a]	Variance	Toluene/Ethylbenzene Mean[a]	Variance	Benzene/Toluene Mean[a]	Variance
0·4	1·137	2·50	1·061	2·04	1·091	1·18
0·5	1·305	1·08	1·097	1·02	1·190	0·84

[a] Mean of 10 results

This shows considerable improvement on previous results, the spread being down to the level given by the flame detector. Further increase in pressure gave no further improvement and gave increased tailing of the peaks.

All the mixtures, together with three more giving an increased spread of component concentration, were run giving the results shown in Table 5.

Table 5. Comparison of mass spectrometer and flame ionisation results using optimised conditions.

	A (%)	B (%)	C (%)	D (%)	E (%)	F (%)	G (%)
Benzene	33·32	33·58	22·22	11·08	33·63	44·56	53·49
Toluene	33·36	35·11	44·44	55·58	33·06	22·17	13·09
Ethylbenzene	33·32	33·31	33·34	33·34	33·31	33·27	33·43

Solution	Detector[a]	Benzene/Ethylbenzene Mean[b]	Variance	Toluene/Ethylbenzene Mean	Variance	Benzene/Toluene Mean	Variance
A	M.S.	1·289	0·7	1·101	0·5	1·171	0·6
	F.I.	2·117	0·4	1·431	0·4	1·479	0·4

B	M.S.	1·309	0·8	1·179	0·5	1·111	0·8
	F.I.	2·093	0·7	1·511	0·6	1·385	0·5
C	M.S.	0·956	1·1	1·495	0·9	0·648	0·8
	F.I.	1·552	0·5	0·840	0·7	0·843	0·4
D	M.S.	0·471	1·1	1·823	1·1	0·258	0·9
	F.I.	0·863	0·7	2·327	0·7	0·345	1·0
E	M.S.	1·359	1·1	1·108	1·0	1·227	1·2
	F.I.	2·216	1·0	1·434	0·9	1·345	0·3
F	M.S.	1·817	1·2	0·774	1·1	2·347	0·9
	F.I.	2·573	1·7	0·999	0·6	2·576	1·6
G	M.S.	2·095	1·2	0·469	0·6	4·465	1·2
	F.I.	2·890	0·9	0·601	0·9	4·810	1·0

^{a} M.S.: Mass spectrometer; F.I.: Flame ionisation
^{b} Means are of 10 results from each solution

Variance figures for both detectors were now of the same order and little further improvement could be expected.

To assess the accuracy obtainable, concentrations of benzene and toluene in the mixtures were calculated from these figures. Mixture E was used as a calibration standard to calculate the sensitivity factors and ethylbenzene (kept constant in concentration for this purpose) as an internal standard. Results are given in Table 6.

Table 6. Composition of mixtures. Solution E used as reference and ethylbenzene as internal standard.

Solution	Component	Mass spectrometer (%)	Theoretical (%)	Flame ionisation (%)
A	Benzene	31·91	33·32	32·15
	Toluene	32·86	33·36	32·99
	Ethylbenzene		33·32	
B	Benzene	32·40	31·58	31·73
	Toluene	35·18	35·11	34·83
	Ethylbenzene		33·31	
C	Benzene	23·68	22·22	23·58
	Toluene	44·05	44·44	42·45

	Ethylbenzene		33·34	
D	Benzene	11·67	11·08	12·20
	Toluene	54·45	55·58	53·69
	Ethylbenzene		33·34	
F	Benzene	44·92	44·56	39·00
	Toluene	23·13	22·17	23·00
	Ethylbenzene		33·27	
G	Benzene	52·04	53·48	44·01
	Toluene	14·05	13·09	13·90
	Ethylbenzene		33·43	

Results obtained using the mass spectrometer as a detector are in most cases as good as those given by flame ionisation and in some cases considerably better. This is particularly so for those mixtures containing a high proportion of benzene which, since it produces the first and sharpest peak in the chromatogram, reaches a maximum concentration in the carrier gas above the linear range of the flame detector.

Linear range of the mass spectrometer detector

In order to determine the linear range of the detector, three dilutions (1 : 10, 1 : 100 and 1 : 1000) of each of nine compounds of a variety of chemical types were made up using ethylbenzene as diluent. This enabled a large range of sample sizes to be injected into the chromatograph, the mass spectrometer sensitivity being adjusted to give peaks within the recorder range in each case. The heights were then adjusted to give those which would have theoretically been obtained using a single sensitivity and were plotted on a log/log scale against sample size. This type of plot gives a straight line (of unit slope) over the linear range of the detector. Absolutely perfect linearity is rarely achieved with any detector over a range of sample compounds. Those giving slopes between 0·95 and 1·05 are generally considered satisfactory.

The compounds used, the m/e values monitored and the results obtained are shown in Fig. 2. The chromatographic column used contained 20% dinonylphthalate on Celite at 100°C. Unit sample size on all the graphs corresponds to 0·1 nanolitre of the pure material injected onto the column. The amount actually detected by the mass spectrometer was of course less than this since some was diverted to the flame detector and some was lost through the intermediate pump. The minimum sample size used was determined by the noise level on the mass spectrometer recorder. 0·1 nanolitre was sufficient to give a satisfactory peak in all cases except that of 1,4-dioxane which required ten times this amount because of its low sensitivity.

The graphs show, for all the materials examined, a slope within the acceptable linear range. The linear dynamic range is approaching 10^4 for all compounds with the exception of dioxane. This compares favourably with many general detectors and is better than has been obtained with other selective detectors. The thermionic detector has a linear range of approximately 1000 and the electron capture detector usually of the

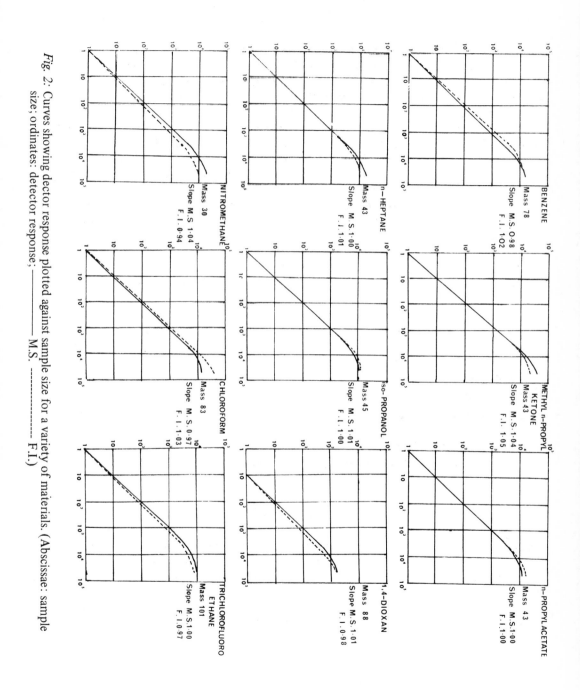

Fig. 2: Curves showing dector response plotted against sample size for a variety of materials. (Abscissae: sample size; ordinates: detector response; —— M.S. ------ F.I.)

order of 500. The curve for benzene shows that with the split used the mass spectrometer is linear for samples about five times as large as the upper limit for the flame detector, as would be expected from the results discussed in the previous section.

Estimation of a minor component eluted on the tail of a major one

In order to assess the performance of the system in practice, an attempt was made to estimate a small amount of toluene which was eluted from the 10% dinonylphthalate column on the tail of a peak due to methyl isobutyl ketone present at four times the concentration.

Choice of mass peaks to be monitored was easy in this case, since methyl isobutyl ketone has a base peak at m/e 43 and gives no contribution to m/e 91. Both toluene and p-xylene, which were used as internal standards, have base peaks at m/e 91 and contribute less than 2% to m/e 43. The peak due to p-xylene was well separated from those of the two components of the mixture.

Results obtained by calculating, both with the use of the internal standard and by the internal normalisation method, are compared with the known composition of the mixture in Table 7.

Table 7. Estimation of toluene eluted on the tail of a large methyl isobutyl ketone peak[a]

	Methyl isobutyl ketone (%)	Toluene (%)	p-Xylene (%)
Theoretical	52·85	13·17	33·98
Using internal standard	52·33	13·30	
By internal normalisation	52·53	13·35	34·12

[a] p-Xylene used as internal standard
Methyl isobutyl ketone: m/e 43, toluene and p-xylene: m/e 91

Estimation of two unseparated components using ions common to both spectra

In the chromatographic analysis of a petrol fraction using a column containing 10% dinonylphthalate, 2,2,4-trimethylpentane (TMP) and n-heptane were eluted together. 2,2,4-TMP has a base peak at m/e 57 and a contribution of about 20% to m/e 43 whereas n-heptane has a base peak at m/e 43 and a contribution of about 50% to m/e 57. Since these ratios are reasonably different it was decided to attempt to estimate these materials using a relationship, the derivation of which has been discussed in a previous paper.[2] The relative concentration of the two components is given by

$$\frac{C_A}{C_B} = \frac{r_B - r_A r_B r_{50} + r_A r_{50} r_S - r_S}{r_B - r_A r_B r_S + r_A r_{50} r_S - r_{50}}$$

where

r_A = height of peak from m/e 57 scanner divided by height of peak from m/e 43 scanner after injection of pure n-heptane.

r_B = height of peak from m/e 43 scanner divided by height of peak from m/e 57 scanner after injection of pure 2,2,4-TMP.

r_{50} = height of peak from m/e 57 scanner divided by height of peak from m/e 43 scanner after injection of a 50/50 mixture of both components.

r_S = height of peak from m/e 57 scanner divided by height of peak from m/e 43 scanner after injection of sample.

C_A = concentration of n-heptane.
C_B = concentration of 2,2,4-trimethylpentane.

The results obtained are shown in Table 8.

Table 8. Estimation of an unseparated two-component mixture using two m/e peaks, each present in the spectra of both components[a]

Component	Theoretical %	Mean[b] %	Calculated Variance
2,2,4-trimethylpentane (isooctane)	55	55.44	0.7
n-heptane	45	44.56	0.8

[a] m/e 43 and 57
[b] Mean of 6 results

Rapid analyses using separate scanners for each component of a mixture

Since separate scanners, each capable of individual gain adjustment, can be used for mixtures of up to six components, the time-of-flight mass spectrometer gas chromatograph combination provides a means of rapid rough analysis of a number of samples of similar composition. Using a standard size sample, the pure ingredients can be injected in turn and the appropriate analogue scanner set to give a full scale reading on the recorder. Within the limits to which the sample size can be reproduced, the percentage concentration of each component can be read directly from the recorder scale.

Three scanners, each producing a trace covering one third of the recorder scale, were adjusted to give full-scale readings for pure acetone, benzene and toluene respectively. Three mixtures of these ingredients made up roughly to the proportions stated, gave for eight successive injections, the scale readings quoted in Table 9.

Conclusions

Provided that care is taken to ensure that the mass spectrometer electronics and all pumping systems are functioning in a stable manner and that dead volume and cold spots in the interface system are kept to a minimum, then a time-of-flight mass spectrometer can be used as an extremely effective selective detector for gas chromatography. It has a good linear dynamic range and does not give anomalous reponses with any of the com-

Table 9.

Component	Mixture composition	Peak height in scale units
Acetone	40%	41, 41, 40, 38, 43, 43, 43, 42
Benzene	30%	29, 30, 29, 28, 31, 31, 31, 31
Toluene	30%	30, 30, 30, 30, 32, 33, 32, 33
Acetone	40%	41, 40, 36, 42, 42, 41, 43, 42
Benzene		0, 0, 0, 0, 0, 0, 0, 0
Toluene	60%	58, 60, 54, 62, 60, 60, 60, 61
Acetone	10%	10, 10, 10, 10, 10, 10, 10, 10
Benzene	60%	56, 56, 56, 56, 57, 56, 56, 54
Toluene	30%	33, 32, 32, 32, 33, 32, 34, 32

pounds examined, such as are found with a number of detectors. It has a wide field of application, can be adjusted in many cases to respond only to one class of compounds in a mixture and gives good quantitative accuracy and reproducibility in the analysis of unresolved components.

References
1   D. Henneburg and G. Schomburg, *Gas Chromatography*, (Ed. M. van Swaay), Butterworths, London, 1963, pp. 191-202.
2   L. P. Lindeman and J. L. Annis, *Anal. Chem.* 32, 1742 (1960).
3   C. C. Sweeley, W. H. Elliott, L. Fries and R. Rhyage, *Anal. Chem.* 38, 1549 (1966).
4   C. D'Oyly-Watkins, D. E. Hillman and D. E. Winsor, *Time-of-Flight Mass Spectrometry* (Eds. D. Price and J. E. Williams), Pergamon Press, Oxford, 1969.

Discussion
Dr. P. F. Knewstubb: The beneficial effects of increasing the intermediate sampling gas pressure from 0.4 torr to 0.5 torr is quite remarkable. Could this improvement in reproducibility be due to the choking-out of variations in the pumping speed?

C. D'Oyly Watkins: I agree that closing the pumping valve does, in fact, reduce pumping speed variations and could be responsible for the improvement. A later sampling system, in which the original large bore valve has been replaced by a needle valve, has given no similar problem.

Chapter 12

The Identification of Cross-Linking Agents in Some Epoxy Resin Systems by Time-of-Flight Mass Spectrometry

C. D'Oyly Watkins and D. E. Winsor

Chemical Inspectorate, Woolwich, England

One of the many functions of an inspection department is the investigation into the causes of failure of equipment in use. Many materials are held in stores for considerable periods of time before being called into use, and hence these failures may occur long after records of manufacturing details have been lost. In this event, chemical analysis can indicate incorrect starting materials or quantities and hence explain the failure.

Modern manufacturing techniques are making increasing use of plastic components. As the properties of the plastic depend very largely on the curing agent used, the ability to recognise this curing agent sometime after manufacture is desirable. Many polymer systems are available and the small number of curing agents and one linear polymer described herein only scratch the surface of a vast field of investigation. These systems consist of the result of crosslinking the diglycidyl ether of bis-phenol A (DGEBA) with some amines and anhydrides - a few of the epoxy range.

Samples were prepared in the laboratory by mixing the polymer and curing agents in the recommended proportions, and then curing the mixture in an oven at the recommended temperature. When cool, the resultant resin was reduced to a fine powder to facilitate filling the sample crucible and to avoid heterogeneity in the sample. One of the problems involved in the analysis of solids using the Bendix 843A Solid Sample Inlet System is the very small amount of sample required in the crucible. The latter is a quartz tube 2½ mm long and ½-1 mm internal diameter. This tube is pinched in the centre so that the sample occupies the top half whilst a thermocouple is enclosed in the bottom. The crucible is heated by a surrounding helical filament after the probe holding the crucible has passed through a vacuum lock into the ion source region. The mass spectra to be shown were recorded at the lowest possible crucible temperature consistent with a full scale chart reading of the highest peak. No spatial significance is necessarily implied in the structural formulae used to illustrate the possible reaction routes and spectra.

Figure 1 shows the general structure of DGEBA and its spectrum, as obtained in a similar fashion to that used for the resins, as shown in Fig. 2. The self polymers are produced by reacting 4,4′-dihydroxydiphenylmethane with epichlorhydrin. By varying the conditions and the catalyst, desired values of n can be produced. This value of n determines to some extent the properties of the polymers - if $n = 1$ or 0, a liquid results, if $n > 1$, a brittle solid. This compound has two sites of reaction, the terminal epoxy group

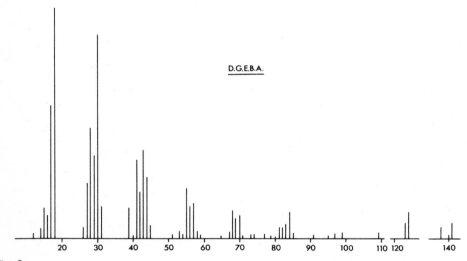

Fig. 1. Poly-DEBPA.

Fig. 2.

and the chain hydroxyl. It is suggested by a number of authors that the amino groups in the curing agents first react with the epoxy bridges (Fig. 3) to form amines which then react with further terminal groups to form crosslinks. If this were a diamine, the result could be doubled above and below the axis of symmetry. Anhydrides, on the other hand, first react with the chain hydroxyl (Fig. 4) to open the anhydride ring, forming an ester which can then react with an epoxy bridge to form a crosslink and another hydroxyl for further reaction. This description is a greatly over-simplified picture of a very large number of possible reaction routes leading to highly complex three-dimensional structures.

Figure 5 shows line spectra of nadic maleic anhydride (NMA) and of the resin it forms with DGEBA. Note the somewhat similar appearance up to m/e 80 and the absence of any significant peaks higher up in the resin spectrum. This appears consistent with the loss and break down of the methylcyclopentadiene ring: NMA is also known as the methylcyclopentadiene adduct of maleic anhydride. A second anhydride cured resin was tried

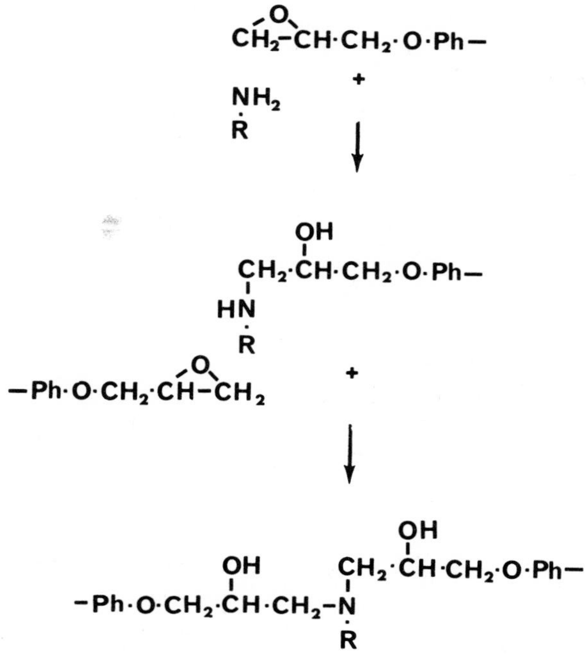

Fig. 3.

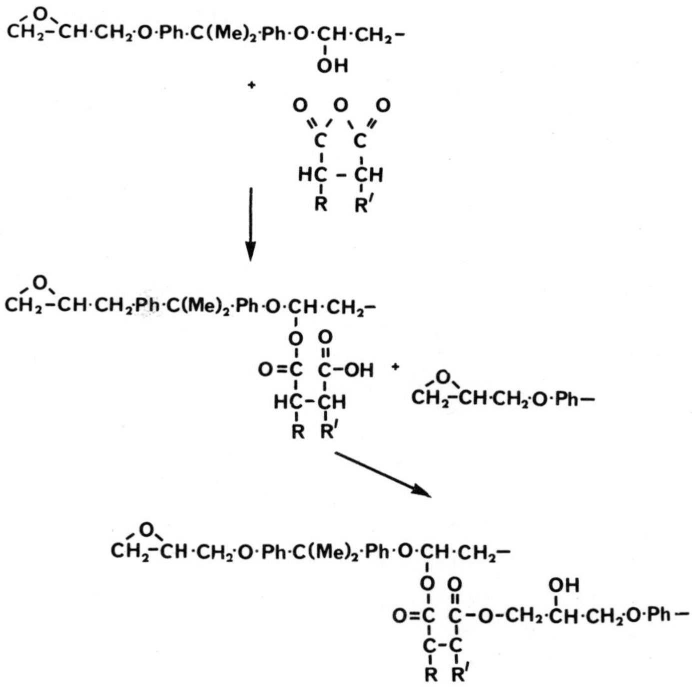

Fig. 4.

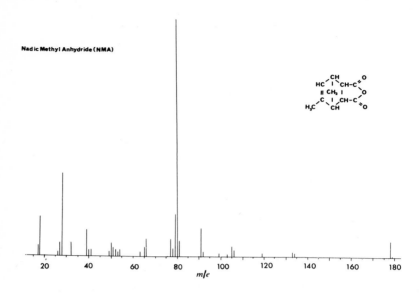

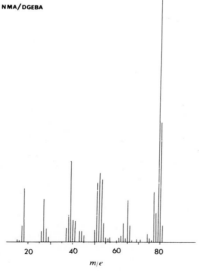

Fig. 5.

The Identification of Cross-Linking Agents in Some Epoxy Resin Systems

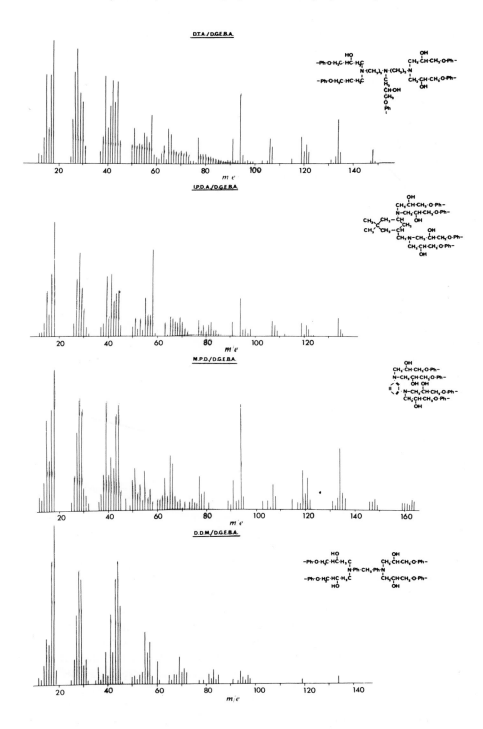

Fig. 6.

but unfortunately it gave an identical spectrum to that of the curing agent alone, even though the system was left heating under vacuum for a long time. More work is necessary to determine whether this is a genuine result or the mixture remains unreacted. It may be significant that both anhydride resins yielded spectra at 100-150°C whereas the amine-cured resins needed heating to 400-500°C.

Figure 6 shows spectra obtained from a variety of such amine-cured resins:
- (i) a straight chain aliphatic
- (ii) a cyclic aliphatic
- (iii) an aromatic
- (iv) an aromatic substituted aliphatic

These spectra are different from each other and from the previous ones and can be used for their identification. They also have certain m/e peaks in common:
- (i) all have a large m/e 28 which is not associated with m/e 32 and is therefore not due to air
- (ii) 55, 56, 57, 58
- (iii) 65, 66
- (iv) 91, 94
- (v) 119, 121
- (vi) 134

In fact, if the three largest peaks in each group of the series 30-49 amu, 50-79 amu, etc. are listed as is shown in Table 1, all four amine-cured resins show three points of coincidence, suggesting an indication of type in an unknown spectrum.

During the course of this work, trouble was experienced with the crucible heaters in the direct inlet probe. These would not stand up to the use (or abuse) to which they were subjected. About this time, Bendix were also having difficulties with their suppliers and considerable delay was experienced in obtaining replacements. In view of this an attempt was made to produce a more robust article. The main problems which had to be overcome were:
- (i) the embrittlement of the heater after some usage and the mechanical damage to it when the quartz crucible was inserted;
- (ii) the possibility of the heating coil sagging during the heating period and shorting onto the filament shields.

In conjunction with our Physics and Glass-blowing Sections, an attempt was made to coat the outside of a quartz tube of slightly large internal diameter than the external diameter of the crucible with a metal conducting film. The first possibility investigated was that of gold. This can be purchased as a liquid to be painted and fired onto the quartz surface to give a film having the correct electrical resistance to be used to replace the heater. Two limitations were experienced with these heaters:
- (i) it is difficult to provide contacts of the same thermal expansion which will maintain electrical contact over a large range of temperature. There is also some evidence for a change of state providing a less adherent film at high temperatures.
- (ii) it was found that at temperatures in excess of about 500°C the film actually evaporated from the quartz surface until insufficient was left to provide an electrically conducting path.

Our latest attempt uses platinum. This can be bought in paste form especially prepared for firing onto quartz. Contacts are easily arranged by attaching platinum wire with the aid of the paste, and the correct resistance can be achieved by laying down successive coats or by rubbing an excess off with abrasive. Only one disadvantage has been apparent to date, that is the quartz tube presents an initial thermal time lag. This is quickly overcome and the temperature then rises smoothly with the increasing current.

Using this type of heater, brittle windings are avoided and sagging onto the filament shield is impossible. One of these platinum heaters has been heated in vacuum for 3½ hours at 700°C without any apparent deterioration.

Table 1.

Polymer	NMA/P	DDSA/P	DTA/P	IPDA/P	MPD/P	DDM/P
43	38	41	39	41	39	44
41	37	43	42	39	44	43
44	41	39	44	44	43	45
55	52	55	58	58	65	55
57	53	69	51	55	66	57
56	51	57	65	57	51	56
84	79	83	77	77	77	71
83	80	71	70	82	79	83
82	77	81	72	81	78	72
99		97	94	94	94	94 *
109		95	91	91	91	97
97		109	106	107	107	95
124		111	119	119	119	119 *
123		123	121	121	121	
		125	120	122	120	
141		137	134	134	134	134 *
138		138	148	135	135	
		139	135	136	148	
		166				
		167				
		152				

Chapter 13

Laser Pyrolysis of Coal and Related Materials in the Source of a Time-of-Flight Mass Spectrometer

W. K. Joy

British Coal Utilisation Research Association, Leatherhead, Surrey, England

and

B. G. Reuben

Department of Chemistry, University of Surrey, Guildford, Surrey, England

Introduction

The laser provides a method for very rapid heating of small particles of solid materials, and the time-of-flight mass spectrometer allows the identification of the gaseous products of such heating at what is virtually the instant at which they enter the gas phase. Robb and Holden[1] have already described the slow heating of coal near the source of a sector mass spectrometer. Vastola and Pirone[2] have described the heating of small lumps of coal in the source of a time-of-flight mass spectrometer using a ruby laser beam as a source of heat.

In a previous paper from this laboratory Joy, Ladner and Pritchard[3] reported the results obtained by laser heating of finely powdered coals in the source of a Bendix time-of-flight mass spectrometer. Their object was to determine the primary products of very rapid heating of coal, particularly with regard to atoms and radicals.

They concluded, from examination of some model compounds mixed with carbon black, that the products of laser-heating were often contaminated with adsorbed gases of low molecular weight. These compounds are further discussed on pages 185 to 187 of this chapter. It was suspected that the coal spectra also included some peaks due to adsorbed gases.

In spite of the contamination of the primary products by adsorbed gases, the spectra produced were characteristic of the various model compounds. It was therefore decided to apply the techique to polymers containing other groups and linkages. It was hoped that this might throw some light on the origin of fragments in the coal spectrum.

Some substances in which carbon was not present were laser-heated to elucidate the role of contamination by adsorbed gases and attempts were made to eliminate the latter by use of a sample holder which could be preheated to 150°C.

Apparatus

The apparatus is shown schematically in Fig. 1.

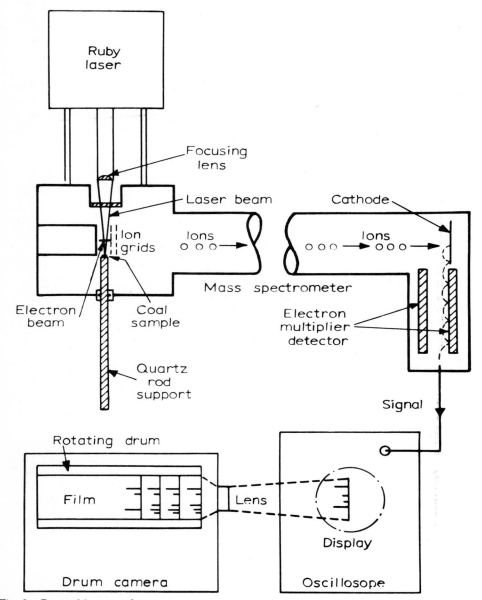

Fig. 1: General layout of apparatus.

A Bendix time-of-flight Model 3015 mass spectrometer was used, with its source modified so that the beam from a ruby laser could be fired vertically downwards through a quartz window in the top flange. The beam was directed between the source grids onto the upper end of a vertical quartz rod which passed through a 'Teflon' seal in the bottom flange of the source. The sample was placed on the upper end of the rod.

The mass spectra were recorded on film by means of a drum camera photographing the oscilloscope display, as described by Kistiakowsky and Kydd.[4] The first recorded spectrum was taken as representing the primary products of pyrolysis. Disappointingly little information could be obtained from subsequent spectra, due to rapid cooling of the sample and the rise of pressure in the ion source.

Samples

Wherever possible, the samples were dusted onto the top of the quartz holder in the form of a fine powder. The coals and the copper powder were applied in this way. White substances such as sugar and ammonium nitrate were pressed onto the rod and the polyethylene glycol, which was a grease, was smeared onto it. 'Polythene' was used in the form of a black sheet, 'Teflon' in the form of a white tape, and the hydrocarbons were melted and a thin film allowed to solidify on the top of the rod.

In the earlier experiments, carbon black was mixed with the light-coloured samples in order that radiation might be absorbed more efficiently. So much gas was desorbed when the carbon black was heated, however, that this had to be abandoned.

Experimental

The sample holder was inserted into the source with the mass spectometer at atmospheric pressure. It was adjusted so that the sample was approximately 1mm below the source grids. The mass spectrometer was then evacuated. Whenever possible, this was done overnight. Immediately before an experiment, the cold traps were filled with liquid nitrogen and when the pressure fell below 10^{-6} torr the electronics were switched on. All runs were performed with an ion accelerating voltage of 3000 V and an electron energy of 70 eV, except in one case where the electron beam was switched off. Unless otherwise stated, runs were carried out at a laser energy of $0.1\ J/mm^2$.

The position of the laser focusing lens was checked and the condensers actuating the flash-tube were charged. During the charging period the drum camera was loaded with film and run up to speed. When the condensers were fully charged, the contacts on the drum camera were closed so that (1) the condensers were discharged through the flash-tube causing the laser ruby to emit a pulse of light onto the sample and (2) the oscilloscope output was triggered and the mass spectra recorded for one revolution of the drum. The time between each spectrum was 0.2 msec and the timing of the flash was such that there were 8 or 9 'background' spectra before the peak of the laser beam energy was reached. Within 0.2 msec of this event, the first spectrum of pyrolysis products appeared.

After the flash, the drum camera was switched off and the film (Eastman Kodak 2475,70 mm recording film) developed at 30°C for 8 minutes in Ilford ID19 developer. The processed films were put into a back projection device, and tracings made of the enlargements.

Results

Polyethylene. (Fig. 2). Polyethylene resembles coal in being a polymer, but is scarcely cross-linked. Figure 2 shows the spectrum obtained from a piece of low density polyethylene sheet into which carbon black had been incorporated. It consists of a series of peaks up to mass 99 separated by 14 mass units. At the time of this experiment, the resolving power of the mass spectrometer was low and assignment of mass numbers above 100 is

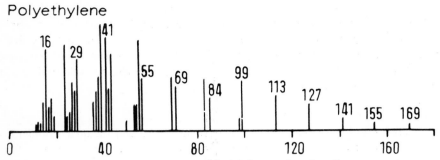

Fig. 2: Mass spectrum from laser heating of 'model' compounds-polymers.

only approximate. Below this, however, fragments correspond principally to the formula C_nH_{2n-1}.

It is difficult to distinguish, both in this spectrum and in subsequent ones, between ions resulting from simple ionisation of pyrolysis products and ions resulting from fragmentation of pyrolysis products under electron bombardment. This will be discussed later.

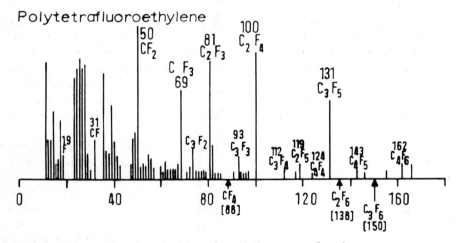

Fig. 3: Mass spectrum from laser heating of 'model' compounds-polymers.

Polytetrafluoroethylene. (Fig. 3). Polytetrafluoroethylene (PTFE) has a similar structure to polyethylene and might be expected to pyrolyse giving fragments differing by units of CF_2 (m/e = 50). Furthermore, as it contains only carbon and fluorine, any hydrocarbon contamination should be clearly visible. PTFE tape dusted with carbon black was used as the sample. Figure 3 shows that contamination is indeed present, and that a series of fragments differing by CF_2 is not obtained though several fluorocarbon peaks are visible.

Pyrene. (Fig. 4). Pyrene is not a polymer, but it is fairly involatile and its structure bears some resemblance to that of coal. The spectrum obtained (Fig. 4) is similar at high mass numbers to the conventional electron-impact mass spectrum of pyrene, suggesting that the sample has merely been vaporised by the laser flash. The peaks at low mass numbers might be due to pyrolysis products but are more likely to arise from contamination.

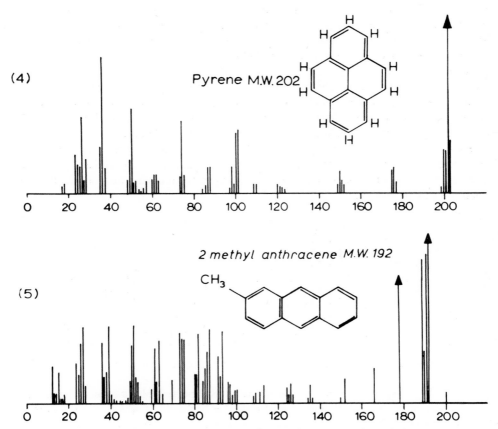

Figs. 4 & 5: Mass spectra from laser heating of 'model' compounds – hydrocarbons.

2-Methylanthracene. (Fig. 5). The remarks made about pyrene apply equally to 2-methylanthracene. The (M–15) peak is prominent in the mass spectrum as would be expected from the mass spectral cracking pattern.

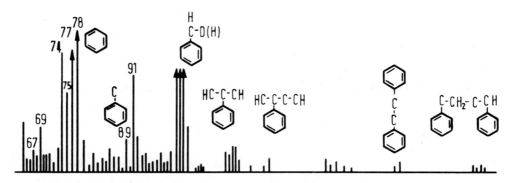

Fig. 6: Mass spectrum of products from the laser heating of polystyrene.

Polystyrene. (Fig. 6). Polystyrene is a linear polymer containing an aromatic nucleus. Figure 6 shows the mass spectrum above $m/e = 64$ obtained by laser heating. The peaks in the region of benzene ($m/e = 78$) and styrene ($m/e = 104$) are off scale. Only small amounts of higher molecular weight ions were found, and probable structures have been suggested in Fig. 6.

Sucrose. (Fig. 7). Sucrose was chosen as an example of an involatile compound which readily undergoes dehydration and thermal decomposition. Its hydrogen-bonded structure makes it almost polymeric in certain respects. The mass spectra of sugars show very large peaks, usually base peaks, at mass 73 corresponding to the stable dihydroxyallyl ion[5] and the presence of this peak in Fig. 7 suggests that some of the sucrose has entered the gas phase. Other peaks in the spectrum are not known to arise from sugars and even allowing for desorption of contaminants it appears that many of them must arise from the thermal decomposition of sucrose.

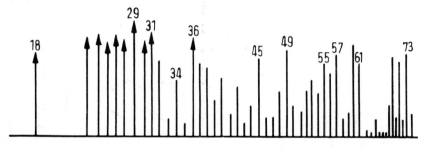

Fig. 7: Mass spectrum of products from the laser heating of cane sugar.

Mass 18 (H_2O) is prominent in the spectrum as expected, plus a series of peaks at masses 24 to 32. Masses 29, 30 and 31 are unusual as background peaks and may correspond to CHO^+, CH_2O^+ and CH_3O^+. There are smaller peaks at most mass numbers which is not surprising in view of the large number of products formed in sugar pyrolysis. The more prominent ones are consistent with the mass spectra of aldehydes which are among the products of such pyrolysis.

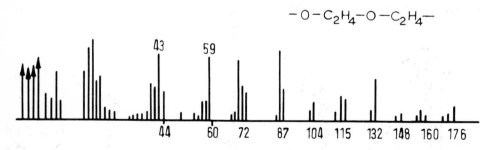

Fig. 8: Mass spectrum of products from the laser heating of polyethylene glycol.

Polyethylene glycol (Fig. 8). The spectrum of polyethylene glycol shows series of high molecular weight peaks. It is possible to obtain three series depending on whether bond fission occurs to give the groups CH_2-CH_2-, $O-CH_2-$ or CH_2-O- at the end of chains.

Laser Pyrolysis of Coal and Related Materials in the Source of a Time-of-Flight Mass Spectrometer

e.g.

m/e =	43	O–CH–CH$_2$	87	CH–CH$_2$–O–CH$_2$–CH$_2$–O
	59	O–CH–CH$_2$–O	104	O–CH$_2$–CH$_2$–O–CH$_2$–CH$_2$–O
	72	O–CH–CH$_2$–O–CH	115	–CH–O–CH$_2$–CH$_2$–O–CH$_2$–CH–O
	70	CH–CH$_2$–O–CH$_2$–CH	132	O–CH$_2$–CH$_2$–O–CH$_2$–CH$_2$–O–CH$_2$–CH$_2$

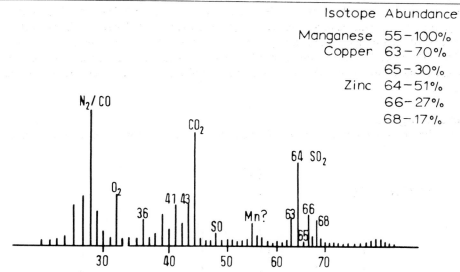

Fig. 9: Mass spectrum of products from the laser heating of copper.

Copper. (Fig. 9). The spectrum obtained on laser heating finely divided copper (<18μm) is illustrated in Fig. 9. It would be expected that if the boiling point of copper were reached (2310°C) there would be a considerable peak at 63 and a smaller one at 65 due to the heavier isotope. It is clear, however, that volatilisation of the copper is not the major source of products. Much of the mass spectrum seems to arise from materials on the metal surface, including hydrocarbon peaks perhaps from adsorbed oil originating in the rotary pump. Together with this, there appears to be some zinc which probably comes from the brass sieves used in sizing the copper powder. The 64 peak may be partly due to

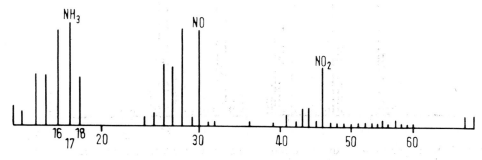

Fig.10: Mass spectrum of products from laser heating of ammonium nitrate.

sulphur dioxide as there is also a small peak at 48 (SO). The peak at 55 could be due to manganese, which has a boiling point of 1900°C and only one isotope. The relatively small peaks at 65 and 63 are probably due to copper which is only slightly vaporised by the laser. Zinc, (b.p. 930°C) of course, has a much higher vapour pressure at all temperatures.
Ammonium nitrate (Fig. 10). Ammonium nitrate was chosen as an example of an inorganic salt because it readily undergoes thermal decomposition and gives gaseous products. Some evidence of hydrocarbon contamination is seen in the mass spectrum (Fig. 10) but the decomposition products are clearly distinguishable i.e. ammonia ($m/e = 17$), water (18), nitrogen (28), nitric oxide (30) and nitrogen dioxide (44). These are the typical products of high temperature heating. The normal product of gentle heating, nitrous oxide ($m/e = 44$), is present only in small amounts, possibly accompanied by some carbon dioxide. The explosive decomposition leading to molecular oxygen does not seem to have occurred.

Adsorbed gases

The observation of quantities of ions of low molecular weights in the spectrum of

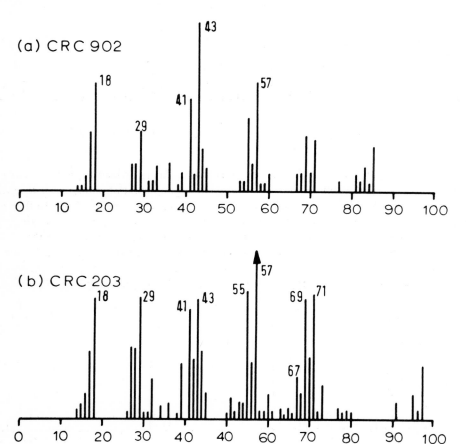

Fig.11: Mass spectra of products from heating coals with a laser beam unfocused and through a 5% transmission filter.

polytetrafluoroethylene which were not associated with carbon and fluorine showed that there was probably contamination on the polymer surface and on the carbon black with which it was mixed in order to increase light absorption.[3]

It was found that the carbon black, when heated by itself, gave off considerable amounts of adsorbed gases so that the practice of mixing it with samples was discontinued. Figures 2, 3, 4, 5 and 8 are the only spectra taken in the presence of this material.

The existence of surface contamination on carbon black and PTFE suggested that similar contamination might have occurred in earlier experiments on coal and is discussed in reference 3. Figure 11 shows spectra of products obtained by heating two coal samples with an unfocused laser through a 5% transmission filter. This should reduce the energy incident on the sample by one or two orders of magnitude. The spectra show hydrocarbon peaks of low molecular weight which again may well have come from pump oil contamination of the surface.

It was decided to try to eliminate this contamination by heating the sample holder and the sample before flashing. A quartz tube of the same outside diameter (8 mm) as the rod hitherto used was sealed and flattened at one end. A small nichrome coil was wound onto a smaller quartz rod, inserted into the open end and brought under the flattened end. A 6 volt supply was found to heat the holder to 150°C in air. Thus the sample holder could be inserted into the 'Teflon' bush and the sample heated *in situ* while the mass spectrometer was being evacuated. The results of heating a CRC 902 coal using a laser 'pumping' voltage of 2100 V (0·1 J/mm^2) are shown in Figs. 12 and 13. The former corresponds to an ionising electron energy in the mass spectrometer of 70 eV while the latter was obtained in the absence of an ionising electron beam, i.e. it shows ions actually produced in the heating process. The second spectrum was recorded at a sensitivity about ten times that of the first.

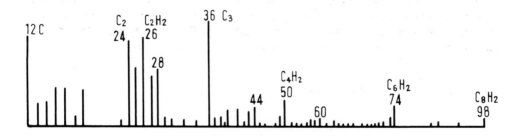

Fig. 12: Coal (CRC902) Heated holder, focused laser, 70 eV mass spectrum of products.

It will be seen that the pattern in Fig. 12 is quite different from that in Fig. 11 and also from the patterns obtained under similar conditions, but without preheating, shown in Figs. 3 and 4 of Reference 3. A large number of peaks have diminished markedly giving a much 'cleaner' spectrum, suggesting that the degassing technique has been successful. It is clear that desorption of adsorbed gases and volatile liquids is one of the major results of laser heating and any attempt to investigate pyrolysis products must take this into account.

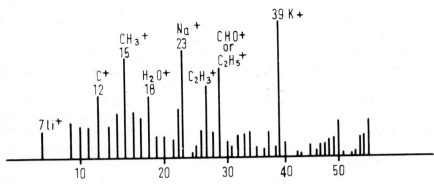

Fig. 13: Coal (CRC902) Heated holder, focused laser, 0 eV. Mass spectrum of products.

Figure 13 shows ions obtained in the absence of an ionising electron beam. These must arise from thermal ionisation processes. Li^+, Na^+ and K^+ were noted in a previous paper;[3] in addition ions at mass numbers 12, 15, 18, 27, 29 and 50 are obtained. These ions were observed only in the first spectrum of the series and had disappeared 200μsec later when the second spectrum was recorded. They correspond, presumably, to C^+, CH_3^+, H_2O^+, $C_2H_3^+$, CHO^+ or $C_2H_5^+$, and $C_4H_2^+$ respectively.

The production of alkali metal ions by thermal ionisation is readily explicable in view of their low ionisation potentials. The other ions, however, have quite high ionisation potentials. Some of these are shown in Table 1 for the production of the ions from their parent molecules or free radicals. This represents the minimum energy required to produce them in the flash heating process e.g. the production of CH_3^+ from methane would require considerably more than 9.9 eV. Ions such as CHO^+ are known to be formed in flames via chemi-ionisation processes, but it would be surprising if such reactions occurred in a laser heating system which does not involve combustion. Nonetheless, the ions must result from a combination of thermal ionisation and chemi-ionisation. It would be interesting to know if the small peaks at masses 19 and 37 corresponded to H_3O^+ and $H_5O_2^+$ both of which are well known in flames and have ionisation potentials below 6 eV.

Table 1. Approximate ionisation potentials.[6]

Parent species	Ionisation Potential (eV)
C	11.3
CH_3	9.9
C_2H_3	9.5
C_2H_5	8.8
H_2O	12.6
CHO	9.9

Effect of laser energy

Three spectra at different laser energies were obtained for the same sample of coal. The CRC 203 coal was chosen as the products were known to be relatively simple.[3] Figure 14 shows the three spectra at energies of 0·025, 0·1 and 0·175 J/mm^2. The peak heights have been corrected for oscilloscope sensitivity and areas of sample irradiated so that there is a direct comparison between peak height and energy per unit area.

Species such as propenyl ($m/e = 41$), acetylene (26) and diacetylene (50) increase

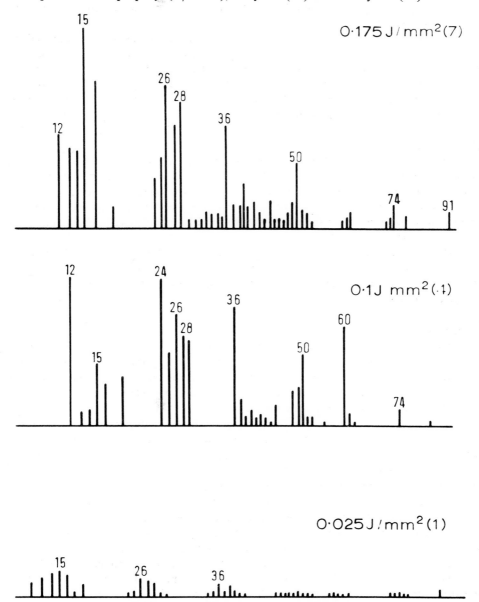

Fig. 14: Mass spectra of laser heated coal (CRC203) at 3 different energies per unit area.

from lowest energy to middle energy but not much more is obtained at the highest energy. On the other hand there is a rapid increase of both methane and methyl radicals. Carbon species, C_1, C_2 etc., are rather erratic. Absent at the lowest energy, they appear to be highest for the medium energy. As they are so short-lived that they only appear on one spectrum, the evaluation of their amounts is difficult because of the lack of synchronisation between the firing of the flash tube and the start of oscilloscope recording. It is interesting to note that free radicals are produced even at the lowest energy (mass 15 is the base peak of the spectrum) and that mass 26 (C_2H_2) is more abundant than mass 28 (CO, C_2H_4 and N_2).

If the low, medium and high energy spectra obtained in this work are compared with those obtained in the previous work[3] where the samples were not preheated, it is seen that the differences are most marked at low laser energies but decrease at higher energies. This is consistent with the hypothesis that at low energies the majority of the spectrum is due to desorbed gases, whereas at high energies, pyrolysis products predominate, with the desorbed gases merely increasing the level of 'background'.

Measurement of temperature

Some time was devoted to attempts to measure the temperature to which the solid particles were being flash-heated. Calculations, assuming no conduction of heat to the sample support, gave obviously absurd values of tens of thousands of degrees. Another attempt, using a two-colour pyrometer[7] to observe the black body temperature of the sample, failed because the scattered red light from the laser could not be elminated by any available system of filters.

It is clear from the results shown in Fig. 9 that using a laser energy of $0.1\ J/mm^{-2}$ a temperature of $2310°C$ (b.p. copper) was not reached. The copper particles, however, showed signs of melting, giving a lower limit of $1083°C$. Further work would be required to bracket this more closely.

Summary

The effects of very rapid laser heating of solid materials have been observed in a time-of-flight mass spectrometer. Many of the earlier results appear to have been clouded by the presence of adsorbed gases on the surface of the solid, and at low laser energies these may be the major products of flashing. It was found possible very much to reduce this effect by preheating some samples *in vacuo*.

Another difficulty in interpreting spectra arose from the uncertainty as to whether a given ion results from ionisation of a decomposition fragment from the heated material, or by fragmentation of a parent molecule by electron bombardment. These two possibilities are well illustrated by pyrene and 2-methylanthracene, which give spectra almost identical with those obtained in conventional mass spectrometers, and ammonium nitrate where the spectrum is almost entirely due to decomposition products.

The spectra of pyrene and 2-methylanthracene suggest that these materials are simply volatilised by the laser. In the cases of coal, polymers, ammonium nitrate and sucrose this was not so and chemical bonds were invariably broken. At all laser energies with coal some free radicals were detected when the heated sample holder was used.

The spectra of coal were similar at higher laser energies to those obtained in previous work,[3] but the degassing procedure showed more clearly that the major decomposition products were C_1, C_2 and C_3 with the polyacetylenes (C_2H_2, C_4H_2, C_6H_2,

C_8H_2), plus methyl radicals, methane,[5] carbon monoxide and dioxide, and the sodium and potassium ions already reported.

The thermal degradation of polyethylene and polytetrafluoroethylene during vacuum evaporation has been studied mass spectrometrically by Luff and White.[9] They found that polyethylene broke down to give an homologous series of hydrocarbons whose carbon chain length varied over a wide range but with a most likely value at three units. The most prominent ion in the mass spectrum was $C_3H_7^+$. The PTFE spectrum on the other hand showed no peaks due to fragments containing more than three carbon atoms. The main peaks corresponded to CF^+, CF_2^+, $C_2F_3^+$ and $C_2F_4^+$ and there was some degradation to elementary carbon.

These workers[9] heated their samples extremely slowly and kept them at a considerable distance from the mass spectrometer ion source. The background pressure was generally about 10^{-5} torr and therefore there seems little likelihood that any free radicals obtained during pyrolysis would survive to be detected by the mass spectrometer. It is almost certain, therefore, that many of the ions which they obtained (e.g. $C_3H_7^+$, CF^+) were due to fragmentation of a stable molecule by electron bombardment.

It is appropriate to compare the spectra obtained here with those of Luff and White, since if laser heating gives rise to transient species, these should show up in one set of spectra but not in the other. In fact, the PTFE spectrum in Fig. 3 is remarkably similar to that of Luff and White and the sequence of peaks CF^+, CF_2^+, CF_3^+, $C_2F_3^+$, $C_2F_4^+$, $C_3F_5^+$ is observed in both sets of data. The greater irregularity of the PTFE spectrum compared with polyethylene may reflect its greater stability and the larger amount of energy required to fragment it.

The polyethylene spectrum shown in Fig. 2 is similar in some respects to that of Luff and White, but exhibits fewer peaks. In particular, the base peak is 39 ($C_3H_3^+$) [or 41 ($C_3H_5^+$) if the former is thought to have been spuriously augmented by potassium contamination.] The alkyl peak, $C_3H_7^+$ is considerably smaller, as is the butyl peak $C_4H_9^+$ and in general the C_nH_{2n+2} and C_nH_{2n+1} peaks are smaller than in the previous work. Mainly C_nH_{2n-1} peaks are obtained together with a moderate yield of methane.

It is probable, therefore, that polyethylene and PTFE are thermally cracked by the laser beam and that the mass spectrum consists of the mass spectral fragmentation pattern of the stable neutral products of the cracking. The differences between the polyethylene spectra obtained here and by Luff and White may be attributed to the greater relative stability of alkenes at the higher temperatures resulting from the use of a laser. Alkane and alkyl ions are therefore relatively less abundant.

The question then arises as to the approximate molecular weight of the stable products of laser heating. The polyethylene spectrum shows that at least a proportion of molecules with molecular weight above 169 must be formed, but these are minor products and the major ones appear to be methane and C_3 and C_4 olefines. This hypothesis would account for most of the large peaks in the spectrum and would permit an analogy with the similar products obtained from industrial thermal and catalytic cracking of n-paraffins. In structural terms, the latter can be thought of as low molecular weight polyethylenes.

Further evidence that the main pyrolysis products are small molecules comes from the polystyrene spectrum. Any high molecular weight pyrolysis fragments from polystyrene would contain aromatic rings and their mass spectra would therefore be expected to show large parent peaks. The absence of such peaks confirms the tendency of laser heating of polymers to give small molecules. The degradation pattern of polystyrene is in fact intermediate between those of polyethylene and PTFE. The hydrogen atoms in the system, as in many of the other systems investigated, appear fairly labile and benzene and styrene were major products. Peaks at 144 and 178 were found which implies some rearrangement on fragmentation of the polymer chain, but these products were present only in very small amounts.

The results with polyethylene glycol were also consistent with the above picture in that the chains fragmented in a fairly random fashion and most of the expected fragments were detected. There appear to be some occasions on which a hydrogen atom is lost to give an ion of odd mass number but why this should happen in only a fraction of cases is not clear.

Conclusions

Laser heating combined with time-of-flight mass spectrometry provides a method for extremely rapid heating of solid materials and examination of products. It is in some ways analogous to flash photolysis of gases, though at present it is in a more primitive stage of development. Stable volatile materials such as pyrene and 2-methylanthracene simply evaporate when laser-heated; thermally unstable materials such as ammonium nitrate decompose. Involatile materials such as coal and polymers are rapidly pyrolysed giving rise to relatively small stable molecules. These result in complex mass spectra which do not clearly correspond to any single precursor.

Joy, Ladner and Pritchard[3,8] have proved the formation of methyl radicals in the laser pyrolysis of coal. This continuation of their work has not shown free radicals to be produced during the laser pyrolysis of other materials, though it is still possible that they are either formed but do not enter the gas phase, or that they enter the gas phase but are not detected. They may possibly be masked by the mass spectral cracking patterns of stable products, but if so they are not produced in sufficiently large quantities to be unequivocally detected.

Experiments with coal using a heated holder showed that free radicals were produced at all laser energies applied. Further work is required to elucidate why free radicals should be produced from coal but not, apparently, from other systems which on the face of it would appear more likely to give rise to them.

Acknowledgements

This work was part of a collaborative research project between the British Coal Utilisation Research Association and the University of Surrey Chemistry Department. The authors thank the Science Research Council for partial support.

Thanks are also due to R. L. Bond for supervision of the work on behalf of the B.C.U.R.A. and to J. G. Edwards of the Department of Optical Metrology, National Physical Laboratory, for measuring the laser energies.

References

1　　H. W. Holden and J. C. Robb, *Fuel* **39**, 39 (1960).
2　　F. J. Vastola and A. J. Pirone, Symposium on Pyrolysis Reactions of Fossil Fuels, Am. Chem. Soc. Meeting, Pittsburgh, C53 (1966).
3　　W. K. Joy, W. R. Ladner and E. Pritchard, *Fuel*, to be published.
4　　G. B. Kistiakowsky and P. H. Kydd, *J. Am. Chem. Soc.* **79**, 4825 (1957).

5   S. Novparast and B. G. Reuben, unpublished work.
6   V. I. Vedenyev, L. V. Gurvich, V. N. Kondratyev, V. A. Medvedev and Ye. L. Frankevich, *Bond Energies, Ionisation Potentials and Electron Affinities,* Edward Arnold, London, 1966.
7   P. Yellow, *Brit. Coal Util. Res. Assoc.* Circular No. 292 (1965).
8   W. K. Joy, W. R. Ladner and E. Pritchard, *Nature* **217**, 640 (1968).
9   P. P. Luff and H. White, *Vacuum* **18**, 437 (1968).

Discussion

Dr. R. I. Reed: I would like to ask how you can justify the ion $m/e = 36$ in Fig. 14 as containing any significant contribution of hydrogen chloride?

W. K. Joy: A 36 peak appears at quite low energies of laser heating when carbon species would not be expected to be present. On the other hand there are chlorides in the mineral matter of the coal which could account for the 36 peak. I said that at high energy a small contribution to this peak still comes from HCl.

Chapter 14

Resolution and Sensitivity of Mass Spectrometers

D. C. Damoth

Scientific Instruments Division, Bendix Corporation, Rochester, New York, USA

Definition[1,2] of the terms resolution and sensitivity for mass spectrometers of medium resolution, e.g. 200 - 2,000, has become of increasing importance due to the growing use of this type of instrument as an analytical and research tool. Standardisation of the nomenclature to minimise the current profusion of definitions is highly desirable. Early clarification will materially simplify the communication of research and analytical data.

One of the most important points to remember in standardising these definitions is that there are three different principles[2] of operation employed in the mass spectrometers in current use. Due to the fact that each of these different principles of mass spectrometer operation (i.e. magnetic deflection; r.f. filter (quadrupole) and time-of-flight separation) have different functions of resolution and sensitivity relative to mass number, other parameters must also be borne in mind. To illustrate this point, examination of the resolution of each of these mass spectrometers at 28 amu and at 250 amu will give different figures of merit for the resolution. Similarly, the sensitivity function and other parameters vary in relationship to each other. This has been a primary cause for the difficulty in standardising the definitions of resolution and sensitivity. It is not the purpose of this chapter to try to standardise the nomenclature, but rather to point out some of the different aspects of the three principles of operation that must be considered if the standardised definitions are to be useful.

Resolution

Some of the common factors which affect the resolution of all mass spectrometers are mass number, sample ion energy spread, ion source pressure, sensitivity setting of instrument, power line (mains voltage) stability, mechanical stability of floor, and sample history since instrument was last cleaned. The definition of resolution (R) used here is that R equals the distance (D) between two mass peaks, M and M+1, divided by the width ($W_{10\%}$) of the peak M, 10% above the base, multiplied by the mass number (M) of that peak, i.e.

$$R = \frac{D}{W_{10\%}} M$$

This definition has been found to be the most practical because it is easily determined and relates to actual usage.

The principle of operation of the mass spectrometer inherently causes the curve of resolution versus mass number to have a characteristic shape. At the risk of incurring some legitimate objections, these curves can be generally visualised as shown in Fig. 1.

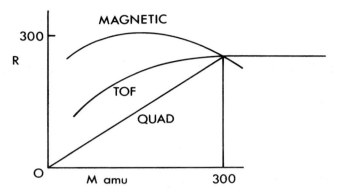

Fig. 1: Curves of resolution versus mass number.

With a nominal resolution of 300, the time-of-flight mass spectrometer would have lower resolution at the low mass numbers due to the peaks becoming sharper than the frequency response of the output system. To achieve a resolution of 300 at mass 28 for such a spectrometer with 3000 volts ion energy and a 100 cm drift path, a frequency response in excess of 10^9 Hz would be required. At higher mass numbers, the resolution improves until the kinetic energy spread of the ions becomes a limiting factor. The curve for T.O.F. resolution in Fig. 1 is typical for an organic molecule fragment ion with an energy spread of about 1 eV. The lack of energy focussing thus limits the resolution of traditional T.O.F. instruments above some mass number, depending on the combination of instrumental parameters and the energy spread of the ions being investigated. This has caused resolution for T.O.F. instruments to be quoted as 200 to 600. The 200 figure is applied for compounds such as perfluorokerosene which have excessive kinetic energy after electron bombardment ionisation and the 600 figure used for compounds such as lead iodide which have very much less excess kinetic energy.

The use of energy focussing in magnetic deflection mass spectrometers is well known. One type of energy focussing in the time-of-flight mass spectrometer was developed over ten years ago and is known as time-lag focussing. Figure 2 illustrates the principles of time-lag focussing. Figure 2A shows the position of the ions at the time of ion formation. The electron beam is then turned off and the ions formed will have motion due to their kinetic energy. The time between the electron beam being turned off and the initiation of the ion draw-out pulse is termed the time lag. When the ion draw-out pulse is applied, the ions are accelerated from the spatial positions as shown in Fig. 2B. Figure 2C shows the focussing action which occurs after the time lag. The ions furthest away from the detector receive additional energy and overtake the frame of reference at the detector. The converse occurs for the ions nearest the detector at the start of acceleration. The net result is that ions of the same m/e ratio reach the detector simultaneously as shown in Fig. 2D. The time lag, which is of the order of microseconds, is easily adjustable for resolution

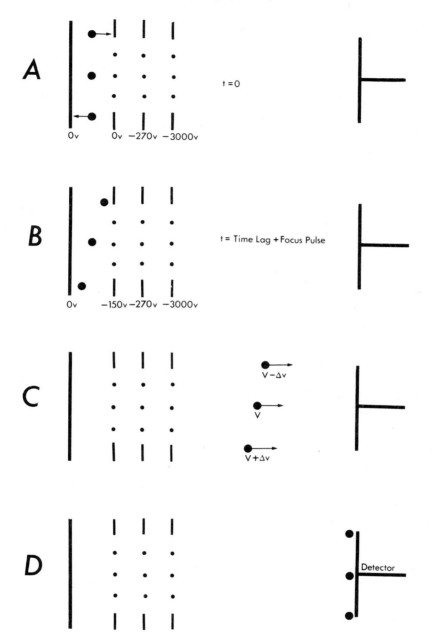

Fig. 2: The principles of time-lag focussing.

optimisation. This energy focussing technique is mass dependent. Mass-dependent focussing is very common in magnetic and quadrupole mass spectrometers. In a sense, the focussing in those instruments is completely mass dependent, except in the case of the Mattauch-Herzog geometry[1, 2], since only one mass peak may be in focus with a given set of parameters. With time-lag energy focussing in the T.O.F. instrument, a condition of maximum energy focus is achieved over a rather broad band of masses, typically 25-50 amu. While the optimum focussing effect is achieved in the chosen mass range, a major portion of the spectrum is visible when displayed on an oscilloscope.

For time-lag focussing to be useful for general analytical purposes, it is desirable to have the position of optimum energy focus coincide with the location, in the mass spectrum, of the T.O.F. analogue output scanning gate. Investigation of the focus action as a function of the mass scale revealed that the position of optimum focus was inversely proportional to the square root of the mass number, $F \propto M^{-1/2}$. This is similar to the function of the mass number with the time-of-flight of the ion between the ion source and the detector, $T \propto M^{1/2}$.

Thus, it was found possible to couple the time-lag focus circuit to the scan drive circuit in a T.O.F. analogue output scanner, permitting the time-lag focus to be optimised at the position in the mass spectrum of the analogue scanner gate. This means that the mass-dependent focussing automatically follows the scanner gate, independently of the scan rate, so that the spectrum recorded by that analogue is energy focussed. This has increased the useable resolution of the T.O.F. instrument for organic species by as much as a factor of two and is particularly useful in organic analytical work since many compounds of interest have molecular weights in the 300 to 600 amu range.

When the improved focussing was employed to increase the resolution, it became evident that other parameters in the T.O.F. instrument must be controlled to achieve resolution of 500 and above. The flatness of the ion grids is a paramount requirement for both good resolution and high sensitivity. A major cause of grid warping was the differential in coefficients of expansion experienced with the molybdenum mesh when sandwiched between stainless steel plates. This effect can now be eliminated by the use of recently developed stainless steel mesh for the ion grids.

Another important factor affecting resolution is the rise time of the ion draw-out pulse, commonly called the focus pulse. Previously it was believed that a rise time of as much as 0.1 to 0.3×10^{-6} sec could be tolerated without affecting the resolution of higher masses. Investigation revealed that a rise time of 0.01 to 0.03×10^{-6} sec, ten times better than previously thought necessary, is required to eliminate this as a significant factor. The improvement of other parameters affecting resolution showed that the slower rise time pulses did indeed cause significant broadening of the lower mass peaks. Careful analysis also showed a noticeable mass peak broadening at higher mass numbers if the slower rise time pulses were used.

The stability of the accelerating voltage is also an important factor. To achieve a resolution of 500 or above, the ion energy must be held constant to within 0.1 V in 3000 V. This necessitates shielding against pickup of a.c. mains voltage by induction, as well as ensuring the requisite stability in the high voltage supply. This effect is particularly noticeable on single-scan observation on the oscilloscope. By watching the oscilloscope trace as single sweeps are triggered manually, the varying positions of the individual isotopes of mercury reveal timing instabilities provided other factors are sufficiently stable.

There is no direct analogy for the trade-off of resolution and sensitivity in the T.O.F. mass spectrometer as exists in a magnetic deflection instrument. In the latter, the slits, particularly the collector slit, may be narrowed to increase resolution while decreasing sensitivity. Similarly, decreasing the bandpass in the quadrupole increases resolution at the expense of sensitivity. In the T.O.F. instrument a small amount of this effect can be obtained by narrowing the slits of the electron gun.

One effect which is more noticeable in magnetic and quadrupole instruments than the T.O.F. instrument is the decrease in resolution as the analyser tube pressure is increased. If space charge effects are not permitted to occur, the resolution loss in the T.O.F. instrument is barely noticeable until the pressure approaches 10^{-4} torr. This is about one decade higher than in a magnetic or quadrupole instrument. The difference is that a collision in the T.O.F. instrument results in the loss of the ion whereas in the others, it causes spreading of the mass peaks.

In the T.O.F. analyser where a straight tube with a uniform potential, typically 3000V, is used as the ion flight path, problems with the flight tube acquiring electrostatic charges which cause loss in resolution do not arise. This is not the case with magnetic deflection instruments, where considerable care must be taken not to introduce samples which will coat the inside of the analyser with a layer which can acquire an electrostatic charge. This is particularly true in double focussing units which employ electrostatic sectors for energy selection. The voltage on these sectors is critical to 0·1 V so that any coating must be avoided.

Quadrupole instruments have a similar but more difficult problem with the coating of the quadrupole rods. This coating is unavoidable in routine analytical use and causes severe and rapid loss of resolution, sensitivity and stability. In quadrupole and magnetic instruments this effect can be reduced by differentially pumping to reduce the pressure in the analyser section. However, such compounds as fluorocarbons should always be avoided. In removing either the electrostatic sector of the double focussing magnetic or the quadrupole's rod assembly for cleaning, care should be used to avoid changing the spacing as this markedly degrades performance. Repositioning the elements is time consuming.

Frequently, the question of resolution is confused with the problem of mass peak identification. The considerable problems with the definitive assignment of mass numbers seems to have led to the circumstance where some researchers use the combination of a large radius, high dispersion magnetic instrument with oil pumps to obtain sufficient resolution to count the individual mass peaks out to the parent molecular ion. This is brought about by the lack of a satisfactory mass marker for magnetic instruments. The hysteresis lag of a magnetic instrument requires several exactly repetitive scan cycles to stabilise. Hall effect probes have been used with some success, but are not yet a satisfactory answer. A similar situation exists with the requirement for a stability approaching 1 part in 10^5 of the electric and r.f. fields in quadrupole instruments to achieve reliable mass marking.

It is possible to calibrate the mass scale of the time-of-flight mass spectrometer by holding the accelerating voltage and other parameters affecting the time-of-flight of the ions constant, while using a highly reproducible time delay circuit and scan generator circuit. In an experimental arrangement, it has been possible to routinely achieve scans from mass 28 to mass 691 reproducible in length by $\pm 0\cdot 2$ amu. Modern electronic digitising equipment is available which can determine scan length far more accurately than strip chart recorders. It is safe to forecast that digitised five-second scans from mass 28 to 700 with unequivocal unit mass assignment at each mass number between will be achievable in the near future at reasonable cost. This will greatly simplify gas chromatograph/mass spectrometer analyses. It is probable that the increased use of output digitising equipment will markedly change resolution definitions for mass spectrometers as most digitising equipment only looks at the tops of mass peaks and their location and not the valleys between the peaks.

Sensitivity

The problem of defining the sensitivity of mass spectrometers is as complex as defining resolution. In this presentation, medium resolution instruments employing electron multiplier detectors will be considered. The matter of comparing such instruments with those employing ion current or photoplate detectors in terms of sensitivity is best disposed of by stating that, in general, instruments with electron multipliers have higher sensitivity. A T.O.F. instrument must have an electron multiplier to be useable.

To make a reasonable basis for examining sensitivity, it is necessary to state what time duration is allowed for the measurement. This is a factor which should not be overlooked in making sensitivity statements. This time dependence of sensitivity is due to the small numbers of ions contributing to a mass peak at very low sample concentrations and is commonly referred to as the statistics problem. Perhaps a greater problem in defining sensitivity is the reference scale. This is indicated by the various units used for measuring sensitivity, e.g. parts per million, torr (mm Hg), grams, divisions of output/micron of inlet pressure, amps per torr. A further complication is that sensitivity varies as a function of mass number at least as drastically as resolution, but not with a 'logical' relationship. For example, any mass spectrometer which uses an electron multiplier detector will lose sensitivity at higher mass numbers due to the relationship $\gamma \propto M^{-1/2}$ where γ is the ion-electron secondary emission ratio at the cathode or first dynode of the electron multiplier.[3] Quadrupoles also lose sensitivity at higher masses due to the decreasing bandpass width necessary to maintain resolution. When this is considered in juxtaposition to Fig. 1, it explains why quadrupoles are not commonly used for gas chromatography/mass spectrometry work or other analytical work in the 300-600 amu range. The sensitivity in that mass range is too low to be of much value when sufficient resolution is achieved.

Another sensitivity factor is the dynamic range, i.e. the maximum signal or pressure which can be present when the minimum detectable output or partial pressure is read. For example, many claims are made of 10^{-14} torr sensitivity with no reference to the maximum ion source pressure which can be tolerated at that sensitivity, nor any other operating parameter at that sensitivity.

Ultimate sensitivity is affected by the residual background. The peaks at every mass number which are observed with oil diffusion pumped systems are helpful for counting masses, but are a hindrance in obtaining maximum sensitivity. Use of rubber gaskets and badly designed sample introduction systems are common examples of instrument design which can lead to an increase in residual background and thus decrease sensitivity. Any useful sensitivity statement must, therefore, take account of the background.

Instrumental factors are frequently quoted in an ambiguous manner to imply sensitivity characteristics. Probably the most common example of this is the reference to the 'duty cycle' of the T.O.F. mass spectrometer. That is, the ratio of electron beam OFF time to ON time. This number ranges from 0.25% where such an instrument has a 250 nsec electron beam pulse width with a 10 kHz spectrum rate, to 5% with a 1 μsec electron beam pulse width with a 50 kHz spectrum rate. This duty cycle is often quoted as evidence that the T.O.F. instrument has less sensitivity than magnetic or quadrupole instruments. This simply is not true. The average trap current or ionising electron beam current of the T.O.F. instrument used for analytical applications is about 10% of the ionising beam current in magnetic instruments in similar applications. The area of the ionisation region is typically 5 mm x 0·5 mm while the source slits in magnetic instruments yield an area of 5 mm x 0·05 mm. This geometry actually yields an ion source output disadvantage for magnetic instruments.

The sensitivity scales of torr partial pressure and A/torr are practically meaningless to analytical chemists. To physical chemists they are of ambiguous meaning until placed in the context of the experimental environment. It is advisable, therefore, that these terms be left to vacuum engineers for use in qualifying residual gas analysers where meaningfull qualifications can be put on their use, or used sparingly in physical chemistry, with preference given to definitions such as particle density or flux in the actual ionisation region.

The units g/sec, ng/sec of a specified compound at a specified scan rate and resolution should be used for dynamic applications such as gas chromatograph/mass spectrometer and direct inlet work. Units such as parts per million of compound X in a matrix of compound Y at a total input rate of N torr/litre or atmospheric cubic centimetres per second at a specified scan rate and resolution could be used for steady state analyses where time is not a critical factor. These are only suggestions which are meant to be helpful in highlighting some problems which are encountered in competitive selling situations.

As noted from the preceding suggested sensitivity statements, the resolution and scan rate are included as standard parts of the definition. This is an essential combination.

The sensitivity of T.O.F. mass spectrometers has been increased by a factor of 10-100 times in the last three years. This has been brought about by (a) increasing the duty cycle, (b) improving the ion transmission percentage and (c) decreasing the noise and drift in the electrometer and electronic amplifier outputs. Factors (a) and (b) have improved the useable sensitivity for both oscilloscope and analogue output. Factor (c) has improved the sensitivity for the analogue output. Related electronic improvements have improved the useable dynamic range for oscilloscope output which has the effect of increasing sensitivity.

The duty cycle can be increased in two ways. First, the electron beam ON time per cycle can be increased. Empirical evidence shows that the maximum practical electron beam duration is about one microsecond. Longer electron beam duration causes increased ion noise, actually yielding a decrease in useable sensitivity in most cases. Secondly, increasing the spectrum repetition rate or number of cycles of operation per second also increases sensitivity. Use of this effect is limited by the decrease in mass range due to overlapping of spectra. Overlapping is a function of drift path length, ion flight velocity and the spectrum repetition rate. Improvements in output system frequency response have allowed the drift tube to be as short as 20 cm for a resolution of 200 at 1,000 V ion energy, permitting a mass range of 1,000 amu with a 50 kHz repetition rate. This is about the practical limit for general analytical use. A repetition rate of 100 kHz is useable for fast reaction studies.

The flat ion grids are the most important factor for maximum ion transmission. It appears that ion transmission can be improved by use of the new stainless steel mesh grids. These avoid warping as previously mentioned, thereby ensuring straight acceleration of the ions towards the detector

The use of solid-state field effect transistor electrometers has become practical since the elimination of premature failures in these devices. This permits the use of all solid-state amplifiers to measure analogue currents down to 10^{-14} A with negligible drift, thus the utmost sensitivity can be obtained when the system is operated at extremely slow scan rates.

The combination of all the above factors has permitted quoting a realistic sensitivity of 1×10^{-9} g of hexane eluted from a gas chromatograph over a time of ten seconds with a scan rate of five seconds at a resolution of 300. If these numbers are not astounding in the face of numbers being quoted or implied in the industry, they are realistic and achievable on a routine basis.

It is noteworthy that much of the productive gas chromatography/mass spectrometry work using magnetic deflection instruments is being performed with very high resolution instruments which have a capability of 30,000 resolution or more but which are operated at a resolution of 300 to 1,000. This is because the dispersion of these instruments is great enough to yield high sensitivity at medium resolution. The same logic dictates that a conventional magnetic deflection instrument with a maximum resolution of about 1,000 will not have extremely high sensitivity when operated at 500-1,000 resolution because the slits will be set quite narrow. Consequently, instruments with resolution of 10,000 are being marketed, not because they are intended to be used at that resolution, which is not high enough to make meaningful mass defect measurements, but rather are intended to be used at about 1,000 resolution to gain satisfactory sensitivity. The main problem with this approach, aside from the cost, is the increased difficulty of fast scanning caused by the heightened hysteresis effects.

Some of the problem of sensitivity loss at higher mass numbers in quadrupoles can be overcome by increasing the size of the rods and the spacing between them. Typical rod size is about 4 mm diameter spaced on about a 1·3 cm circle. This leaves a very small space between the rods. However, increased electronic power is needed to match the increased rod size, so this approach is limited.

A T.O.F. instrument using the time-gated magnetic electron multiplier has the unique ability to discriminate against multiplier noise output at any mass number. This is due to the fact that the analogue gate only opens for about 0·1% of the time of each cycle; just long enough to admit the signal of the mass peak to be scanned or monitored. This 1,000 to 1 advantage, together with the figure of less than 10 noise counts/sec total for the mass (in some cases approaching 0·1 count/sec) yields a noise figure of less than one per 100 sec at any mass number. Usually the random ion noise in the spectrum or the background gases are the limiting factors. Making the instrument bakeable to higher temperatures, such as 350°C, helps to reduce background signals.

Random ion noise in the T.O.F. instrument is brought about primarily because practically all ions produced in the ion source are seen in the output since the transmission losses are not great. This, then, is both one of the most positive virtues and biggest drawbacks to the T.O.F. principle. Magnetic and quadrupole instruments pass only one species at a time, and therefore it is somewhat easier for these instruments to eliminate unwanted masses. In particular, the geometrical separation parameters of magnetic instruments offer a straightforward means to improve the resolution/sensitivity situation by increasing the dispersion and widening the slits, or by other methods which have the same result.

In conclusion, it should be stressed that it is essential to consider the characteristics of the instrument in conjunction with the resolution and sensitivity definition. Prospective mass spectrometer users would be well advised to consider most carefully the characteristics of the operation principle of an instrument and not simply performance claims when selecting an instrument for a particular usage.

References
1   J. H. Beynon, *Mass Spectrometry,* Elsevier, 1960.
2   J. Roboz, *Introduction to Mass Spectrometry, Instrumentation and Techniques,* John Wiley and Sons, 1968.
3   M. Ackerman, F. E. Stafford and J. Drowart, *J. Chem. Phys.* **33**, 1785 (1960).

Comment
Dr. D. W. Thomas: With regard to multiplier discrimination one need only consider the effects in the vicinity of the multiplier first dynode or cathode. There are three main effects, the relative magnitude of which is still undetermined. The first and most widely known effect is the variation of the secondary electron coefficient for ion impact at a metal surface. Such coefficients are functions of many parameters, including the mass, charge, chemical structure, and velocity of the ion, and also the angle of incidence, degree of gas adsorption, and the temperature of the surface. Some general conclusions are that for ions in the energy range 1 to 5 keV:

(a)   The gain is approximately a linear function of energy whatever the nature or chemical structure of the ion. The range of linearity increases as the mass increases.

(b)   At equal energy, the gain varies inversely with the square root of the mass of the isotopes of an element, particularly for $m/e > 40$. For $m/e < 20$, the gain increases as mass increases and in the intermediate range the gain appears independent of mass.

(c)   At equal velocity, the gain increases with atomic number, particularly for $5 < Z < 10$ and more especially as the velocity is increased.

(d)   At equal velocity, the gain measured for polyatomic ions diverge from a common value as the velocity is increased,

and

(e)   At equal velocity, for an element X the gain of X^{2+} is approximately half that of X^+, the numerical factor decreasing as the velocity decreases below 10^7 cm sec^{-1}.

Mahadevan *et al.*[4] found that the secondary electron yields are a function of the state of cleanliness of the cathode, particularly that fraction of the yield due to potential ejection. For masses in the range 1 to 32 amu they found that the yield is linear with energy in the energy range[9] 800 to 2000 eV and that above a certain velocity threshold the yields are proportional to ionic mass at constant velocity. In the energy range 3 to 10 keV, Schram *et al.*[5] found that, for singly and multi-charged rare gas ions, the gain was a linear function of velocity, independent of charge, with the slopes of the lines proportional to the square root of the mass. Thus the gain increases as the mass decreases, for constant ion energy. At higher impact energies (> 10 KeV), Gibbs and Commins[6] found that the order of yields is $K^+ > Na^+ > Li^+$, in agreement with Inghram.[1] For negative halogen ions however Gibbs and Commins found that the reverse order applied. They further noted that the yield was approximately linear with velocity.

C. la Lau has examined the relative gains (compared to the $C_3H_5^+$ (mass 41) ion as unity) of many organic ions covering the mass range 25 to 350 amu. He has interpreted his results, not in terms of the mass of the molecular ion, but in terms of (a) the nature and number of the constituent atoms and (b) the kinetic energy carried by each atom. It has suggested that the molecular ion breaks up on striking the target and acts as a group of particles sharing the kinetic energy. Thus, using further the experimental observation that the yield for monatomic ions increases roughly linearly with kinetic energy, la Lau suggests that the relative yield of an ion of mass M at a certain energy E and containing n_i atoms of atomic mass m_i may be written as $\gamma_r(M) = \frac{1}{M} \Sigma_i n_i m_i \gamma_{ri}$ where γ_{ri} is the relative yield per atom m_i at the same nominal energy E but where the fractional energy carried by m_i is $\frac{m_i}{M} \cdot E$. Using this additivity principle, la Lau's calculations of the relative gains of complex organic ions agreed with the experimental values to within an average deviation of 5%.

The second factor which may give rise to mass discrimination in the Bendix electron multiplier is the geometric mass discrimination at the multiplier. This arises from ions being deflected by the magnetic field so that the various entrance apertures intercept a mass-dependent fraction of the incoming ions and prevent them being detected. This has been investigated by Hunt et al.[7]

The same authors have also done some unpublished work on the third discriminating factor, namely that which arises from the mass-dependent location of the particle impacts on the first dynode due to the magnetic field. The efficiency of focussing these secondary electrons onto the multiplier dynode strip may well vary with their position of origin on the first dynode.

References

1   M. G. Inghram and R. J. Hayden, 'A handbook on mass spectrometry'. *Nucl. Sci. Ser.* No. 14 1954 Natl. Acad. Sci. − Natl. Res. Council 311 pp. 41 to 44.
2   F. Hoffert, J. Paulus, and J. P. Adloff, *Rev. Phys. Appl.* **1**, 43 (1966).
3   M. Kaminsky, *Atomic and Ionic Impact Phenomena on Metal Surfaces,* Springer-Verlag, Berlin, 1965, pp. 300 to 332.
4   P. Mahadevan, G. D. Magnuson, J. K. Layton and C. E. Carlston, *Phys. Rev.* **140**, A1407 (1965).
5   B. L. Schram, A. J. H. Boerboom, W. Kleine and J. Kistemaker, *Physica* **32**, 749 (1966).
6   H. M. Gibbs and E. D. Commins, *Rev. Sci. Instr.* **37**, 1385 (1966).
7   W. W. Hunt, K. E. Megee, J. K. Streeter and S. E. Mangham, *Rev. Sci. Instr.* **39**, 1793 (1968).

Discussion

D. E. Winsor: What is the significance of the filament shields in the direct inlet system?

D. Damoth: The purpose of the shields on the Direct Inlet Probe (Model 843) is to reflect the heat from the heater filament. The unit was designed to heat samples to over 750°C. The power required to raise the temperature of the sample crucible to that level caused too much heat to be radiated to the ion source elements, so the heat shields were added.

C. D'Oyly-Watkins: Could you comment on the fluctuations in the position of the analogue gates which appear to vary from time to time?

D. Damoth: The analogue gate position is affected by changes in the heater temperature of the thermionic tubes, caused by mains voltage changes. Ascertain that the line voltage is constant to $\pm$ 0·5 volts and utilise a constant harmonic output (sinusoidal waveform), time voltage regulator and the problem should be solved. Most people assume their mains voltage is constant, but it practically never is.

J. E. Williams: I think it is generally agreed that the multiplicity of definitions and terms relating to resolution and sensitivity in mass spectrometry is confusing and a serious problem in communicating technical and research data. Your paper has quite rightly highlighted the problems in relating the manufacturer's performance claims for mass spectrometers of widely differing basic principles. It is to be hoped that the appropriate international organisation will be stimulated to seek a precise definition and standardisation of terms, in co-operation with equipment manufacturers and research workers.

In a lighter vein, may I point out that the sensitivity definitions of torr, partial pressure and amps/torr — which you have stated to be 'meaningless to analytical chemists' — are widely used in your own company's literature!

D. Damoth: That is true, but as you know, our mass spectrometers are used by physical chemists and physicists, as well as organic and analytical chemists. These terms are more useful in those areas of research.

Chapter 15

A Linear-to-Logarithmic Compressor Circuit for Electronic Recorders

K. O. Dyson

Department of Chemistry, University of Cambridge

The multiplier output signal from a mass spectrometer may be fed to a pen recorder after suitable amplification. If there is a logarithmic range within the amplifying system this is often the mode selected for the initial scan over the complete mass range of a new sample. The type 321 analogue scanners fitted to the Bendix T.O.F. mass spectrometer at Cambridge University do not have this facility and a simple, inexpensive method was used to achieve the required effect.

Rectifiers of the contact barrier type have a forward impedance, Z_f, which follows the law,

$$E = r \log I \qquad (1)$$

Voltage = Resistance x log Current

over a wide range of the forward characteristic. Copper oxide elements, originally used as signal detectors and meter rectifying diodes, follow this law consistently for a current range of between 1 and 50 above the potential of 0·07 V per contact layer. A simple circuit for logarithmic conversion is the voltage divider shown in Fig. 1a. The current through the rectifier is logarithmically related to the input voltage if:

(i)   the ohmic resistance of the component parts is $\leq Z_f$.
(ii) the source resistance, R_s, $\geq Z_f$.
(iii) the load resistance, R_m, $\geq Z_f$.

All these conditions can be met quite easily.

Figure 1b shows the circuit which was used to test a WM 162 copper rectifier and Fig. 2 shows a plot of E_{in} against E_{out}. E_{in} was obtained from dry batteries and measured by a valve voltmeter; E_{out} was measured as % f.s.d. of a 10 mV recorder. The plot shows that Eqn. (1) is followed over nearly two decades, qualitatively, and over the linear portion of the graph the relation

$$E_{in} = 2 \cdot 3 \, e^{\, 7 \cdot 1 E_{out}}$$

was obtained from the original (E_{in} = applied volts, E_{out} = measured millivolts across 18 kΩ in the test circuit.)

Figure 3 shows the final design which was built into a voltage attenuator circuit used as an interface between the Bendix 14-105 T.O.F. mass spectrometer and a Vitatron

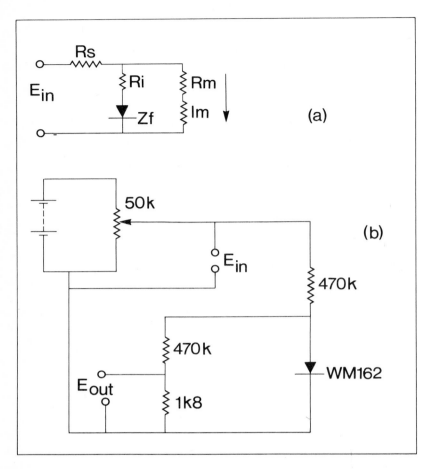

Fig. 1: (a) Simple voltage divider network for logarithmic conversion of E_{in} to I_m.
(b) Test circuit to determine if a selected diode behaved according to the theoretical prediction. The 1K8 was selected to give FSD on the recorder for the maximum E_{in}.

Recorder, type UR100. Provision has been made for selecting the 0-20 V signal from the 321 Scanners and feeding it into a 0-20 V log i/p or 0-20 V linear i/p network. The 0-2 V i/p is for coupling to the Veeco Ionisation Gauge Control Unit; the variable i/p is useful for other signals.

In use the electrometer switches should be set so the largest peak in the mass spectrum is just on scale (i.e. almost 20 V). Then the logarithmic compressor (it is just that and not an amplifier) will boost the smaller signals at the expense of the larger. This will enable the entire spectrum to be scanned on one range setting of the electrometers at a useful sensitivity. Further recordings may then be made of the mass regions of interest in the linear mode.

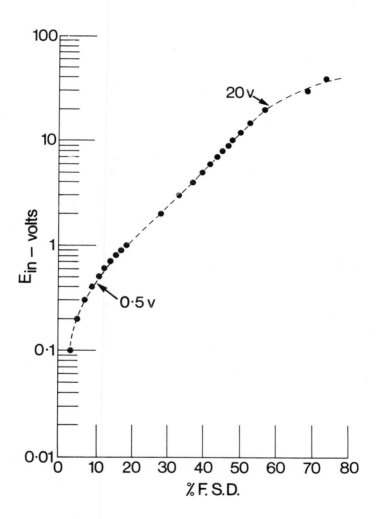

Fig. 2: Plot of E_{in} versus % FSD of 10 mV recorder.

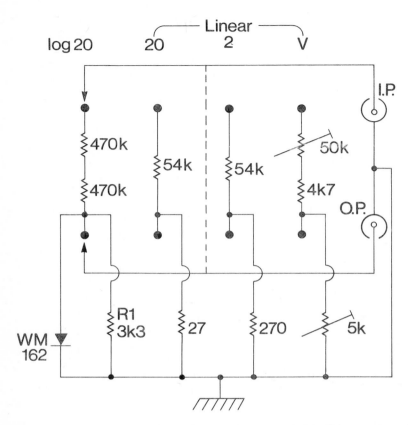

Fig. 3: Voltage attenuator unit with logarithmic compressor circuit. R1 was selected to give FSD on the recorder for 20 V i/p.
Input Z of unit; log $\simeq$ 500k Ω
 lin $\simeq$ 50k Ω
Output Z; varies 27 to 5k Ω
Z_f of diode was 50 k Ω at 1·5V.
Many diodes, besides the contact barrier type, may be used as compressor devices but may not provide a logarithmic conversion.

Chapter 16

An Amplitude Limiting Device for U.V. Galvanometer Recorders

C. D'Oyly-Watkins and S. N. Gaythorpe

Chemical Inspectorate, Woolwich, England

When a mass spectrum is being scanned at high gain, using a u.v. recorder to monitor the output, ions in high abundance can, for short periods, produce currents greater than the maximum safe limit of the recorder galvanometers. If the mass spectrometer is used as a detector to monitor gas chromatographic effluents, such currents can exist for much longer periods. Under such circumstances some form of current limiter, to prevent current flow in excess of that needed for full scale deflection on the recorder, is desirable. Similarly, if more than one channel of output of a time-of-flight mass spectrometer is being used to scan different parts of the spectrum or to produce a number of selective chromatograms simultaneously, limitation of each galvanometer's swing to a separate part of the recorder scale avoids confusion from overlapping traces. In order to provide these facilities, the limiting device described below was constructed.

The design criteria for the instrument were as follows.
1   It should be relatively inexpensive since a separate unit was required for each of the six channels of the galvanometer drive amplifier in use (S.E. Laboratories 425).
2   The cut-off should be as sharp as possible so that the amplitude of peaks below the set limit was unaffected.
3   It should be easily switched to provide limits of full, half and one third scale.

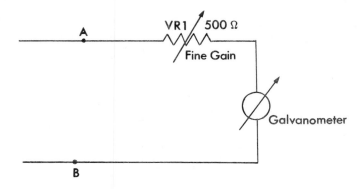

Fig. 1.

The output current from the SE 425 amplifier reaches the galvanometer as shown in Fig. 1, through a fine gain control, a 500 Ω helipot (VR 1). This potentiometer is normally set at about 250 Ω since it also acts as the damping resistor for the galvanometer. With the 75 Ω galvanometers employed (SE type A 1000) the current can be prevented from exceeding the value of 9·86 mA required for full scale deflection, by limiting the voltage across the galvanometer to 0·74 V. Alternatively the voltage across both the galvanometer and VR 1 in series (points A and B in Fig. 1) can be limited to 3·20 V. Use of the latter method has the added advantage that fine adjustment of the cut-off point can then be accomplished by slight alteration in the setting of VR 1.

Zener diodes, normally the obvious choice for such an application, do not function very satisfactorily at the low voltage level needed and would not give a sufficiently sharp cut-off. The alternative possibility of using a normal diode was equally uninviting due to the cost of the six high-stability power supplies which would be needed to provide separate backing-off voltages for each channel. Fortunately, two new types of voltage dependent resistor, possessing characteristics similar to those of zener diodes, have recently become commercially available. The voltage/current characteristic of these 'asymmetric voltage-dependent resistors' (AVDR) is shown in Fig. 2. One of these (Mullard Type E295ZZ/01) has a forward 'breakdown' at 1 V ($\pm$ 10%) and a maximum forward current of 25 mA. The slope for a typical resistor is such that the current increases from 0·1 mA to 10 mA as the applied voltage changes from 0·85 V to 1·15 V.

Three of these resistors were connected across points A and B as shown in Fig. 3. Switch S1a enabled one or two of the AVDR to be shorted out, providing the lower

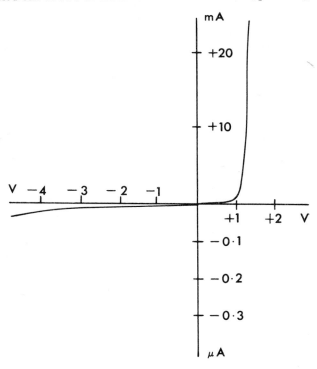

Fig. 2.

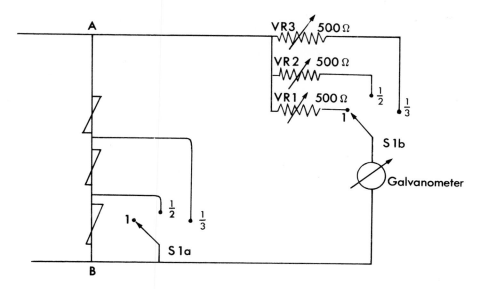

Fig. 3.

limiting voltages for the shorter spans. Potentiometers VR 2 and VR 3 were included to replace VR 1 for these spans and were switched simultaneously by S1b. The three potentiometers were set to provide galvanometer spans of full, half and one third full scale respectively. A plot of galvanometer response against input signal, obtained by increasing the input to the drive amplifier in steps, by adjustment of the scanner attenuator, gave a constant slope until the output was within one or two percent of the cut-off level.

Chapter 17

A Gas Chromatograph/Mass Spectrometer Interface System

A. J. Luchte and D. C. Damoth

Bendix S.I. & E. Division, Rochester, USA

Introduction

During the past decade several different methods of connecting gas chromatographs to a mass spectrometer have been reported (G.C./M.S. systems).

One of the initial methods used, and still widely incorporated in many G.C./M.S. systems, is the use of the splitter valve to reduce the pressure and flow rate of the gas chromatograph (G.C.) to that acceptable by the mass spectrometer.[1] This method, however, uses less than 1% of the G.C. effluent and the resultant sensitivity is low.

A more efficient connection utilises the 'molecular separator' principle, in which the carrier gas concentration is reduced by selective diffusion through some porous material or 'jet-molecular diffusion'.[2] The porous material systems have been constructed of fritted glass tubes,[3] porous Teflon,[4] and porous metal.[5,6] Llewellyn[7] used an opposite approach with a silicone rubber membrane through which the organic molecules permeate and the helium carrier gas is rejected.

Although these separators use widely differing techniques, the resultant transmission of usable sample into the mass spectrometer ion source is essentially the same for packed columns. This is because the maximum amount of sample which can be passed into the mass spectrometer is limited by the amount flowing from the gas chromatograph.

Experimental

In order to provide a complete packaged interfaced system for use with the time-of-flight mass spectrometer and any gas chromatograph, while fulfilling the requirements to be easy to connect and use, low in price yet completely reliable, we decided to use a silver membrane separator.[5,6] Other types were ruled out because of their complexity, poor durability, and high production costs.

The interface is connected in a manner suggested by Grayson and Wolf.[8] This method was found most satisfactory as it eliminates the need for a splitter valve at the exit of the G.C. column, which limits the ultimate sensitivity of the system as part of the sample is thrown away. The use of capillary tubing as a pressure reducer to the separator, as suggested by Grayson and Wolf,[8] was the most convenient method available.

The selection of the silver membrane was made after tests on both it and the glass

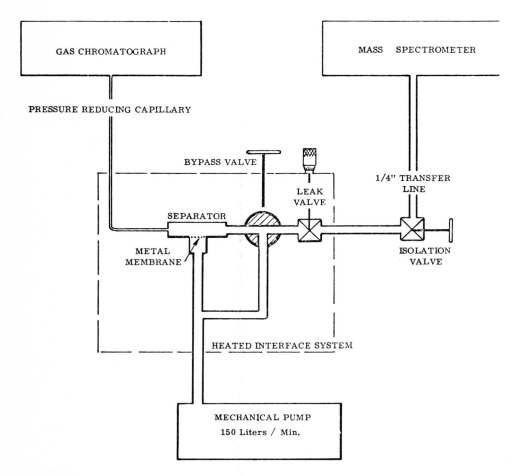

Fig. 1: Bendix GCI-1 gas chromatograph/mass spectrometer interface.

frit type separator. The rigidity of the metal separator made it the favoured unit for production as well as everyday use in the laboratory. As reported by several of the aforementioned authors, the glass frit retains some samples causing peak tailing and memory whereas the metal membrane does not.

Enrichment and yield figures were very close to those reported by Blume,[7] but it was felt there was a need for further definition of an important parameter. That was, 'How much smaller sample size could be passed into the G.C. column and analysed by the mass spectrometer by using the separator than by using an ordinary splitter valve?'

Hexane was used as a test compound with toluene as the solvent. All tests were run using a Bendix Chromo-Lab 2100 Gas Chromatograph interfaced to a Bendix 3012 mass spectrometer. The total output integrator of the mass spectrometer was used as the detector for recording the chromatogram, except when tests of the delay time through this interface and peak distortion were run. At these times the thermal conductivity detector was used together with the total output integrator.

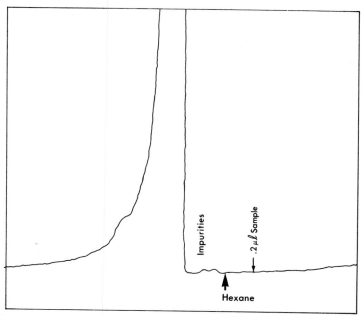

Fig. 2: G.C./M.S. Splitter connection. Sample ratio: 10ml toluene: 1 μl hexane.

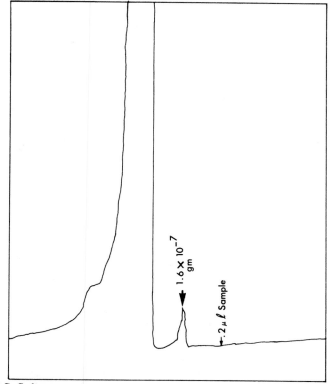

Fig. 3: G.C./M.S. Splitter connection. Sample ratio: 10ml toluene: 10 μl hexane.

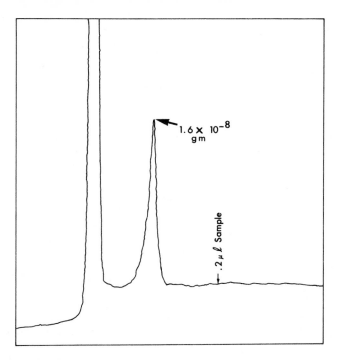

Fig. 4: Bendix GCI-1 Interface. Sample ratio: 10ml toluene: 1 µl hexane.

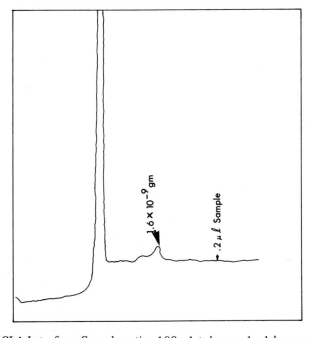

Fig. 5: Bendix GCI-1 Interface. Sample ratio: 100ml toluene: 1 µl hexane.

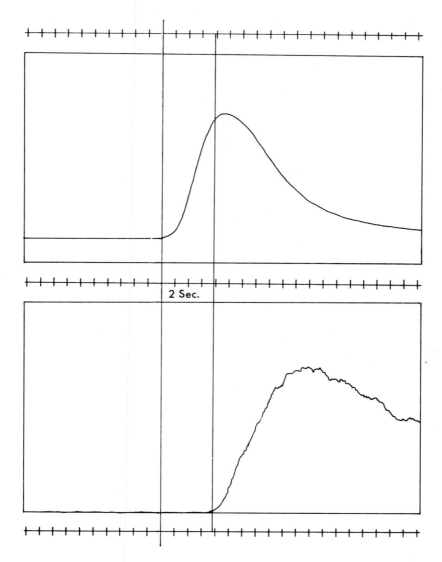

Fig. 6: Delay through separator. Top: Detector; Bottom: MS-TOI.

Figures 2, 3, 4 and 5 show peaks recorded on the Bendix total output integrator for various sample sizes using both a splitter valve and the Bendix interface. Figure 6 shows a time delay of two seconds, that is transmission time through the interface system. Note that almost perfect peak symmetry is maintained.

Flow rates tested ranged from 5 to 55 ml/minute with a separator temperature of 200°C. The smallest sample of hexane inserted into the gas chromatograph and recorded by the mass spectrometer, with the use of a splitter valve, averaged 10^{-6} g with the flow rates mentioned. With the use of the molecular separator interface, the smallest recordable sample was $1 \cdot 5 \times 10^{-9}$ g.

Conclusion

It is considered that this interface system is simple to use, extremely rugged, and very efficient. Although one interface system cannot do all G.C./M.S. work, we believe the Bendix GCI-1 will meet the requirements of most present day anaylsis. We feel it will be an excellent addition to systems incorporating a straight splitter valve type of interface.

References

1. R. S. Gohlke, *Anal. Chem.* **31**, 535 (1959).
2. R. Ryhage, *Anal. Chem.* **36**, 759 (1964).
3. J. T. Watson and K. Bieman, *Anal. Chem.* **37**, 1135 and 8446 (1965).
4. S. Lipsky, C. G. Horvath and W. J. McMurray, *Anal. Chem.* **38**, 1585 (1966).
5. M. Blume, *Anal. Chem.* **40**, 1590 (1968).
6. R. F. Cree, *Pittsburgh Conf. on Anal. Chem., March 1967,* Abstract of Papers, p. 96, No. 188.
7. P. M. Llewellyn and D. P. Littlejohn, *Pittsburgh Conf. on Anal. Chem.* February, 1966.
8. M. A. Grayson and C. J. Wolf, *Anal. Chem.* **39**, 1438 (1967).

Bibliography of Papers Involving the Use of Linear Non-Magnetic Time-of-Flight Mass Spectrometers

W. K. Joy

The British Coal·Utilisation Research Association, Leatherhead, Surrey, England

Introduction

Every effort has been made to include all published papers which are readily available. Use has been made of the *Mass Spectrometry Bulletin* up to August, 1969, and subsequent publications can be traced through this source, which is published by the Mass Spectrometry Data Centre, A.W.R.E., Aldermaston, Berkshire, England. A few early papers have been traced through the *Abstracts and Bibliographies of Articles and Papers on the Bendix Time-of-Flight Mass Spectrometer* published by the Scientific Instruments Division of the Bendix Corporation, 1775 Mount Read Boulevard, Rochester, New York, N.Y. 14603, USA.

The Bibliography is divided according to application. The references have been placed in the most appropriate sections for their content and are in chronological order in each section. In some cases a paper could equally well be placed in one of two categories (e.g. Mass Spectral Data and Gas Chromatograph/Time-of-Flight Mass Spectrometer). Care should, therefore, be taken when using the Bibliography since each paper is quoted once.

My thanks are due to members of staff of the B.C.U.R.A. library and the University of Salford Chemistry Department for assistance in checking some of the references.

1   Bibliography and reviews

Time-of-flight mass spectrometry, Harrington, D. B., *Encyclopedia of Spectroscopy* (Ed. Clark, C. F.), Reinhold, New York, 1960, p. 628.

The time-of-flight mass spectrometer — its applications. Tomita, H., *J. Chem. Soc. Japan* **67**, 1769 (1964).

Le spectrometre de masse à temps de vol et ses applications. Couchet, G., and Zvenigorosky, A., *Rev. Quest. Sci.* **26**, 167 and 334 (1965).

Recent advances in time-of-flight mass spectrometry. Damoth, D. C., *Adv. Anal. Chem. Instrumentation* **4**, 371 (1965).

Advances in time-of-flight mass spectrometry. Damoth, D. C., *Mass Spectrometry* (Ed. R. I. Reed), Academic Press, London, 1965, p. 61.

Review of time-of-flight mass spectrometry. Thirion, B., *Methodes Phys. Anal,* 62 (1966).

Characteristics of time-of-flight mass spectrometry. Hikita, T., Ishihara, F., Sueyasu, S., and Tsuchiya, S., *Mass Spec. Japan* **14**, 157 (1966).

Time-of-flight mass spectrometry. Joy, W. K., *British Coal Utilisation Research Association Monthly Bulletin* **31**, 581 (1967).
The time-of-flight mass spectrometer. Price, D., *Chem. Brit.* 4, June 1968.
Abstracts and Bibliographies on the Bendix Time-of-Flight Mass Spectrometer, Cincinnati Division of the Bendix Corporation.
Bibliography. W. K. Joy, in *Time-of-flight mass spectrometry* (Eds. D. Price and J. Williams), Pergamon Press, Oxford, 1969.
Mass spectrometers (Russian), Rafal'son, A. E. Shereshevskii, A. M. Book. Atomizdat, Moscow, 1968; *Nucl. Sci. Abstr.* **22**, 4384 N. 43160 (1968).
Non-magnetic time-of-flight mass spectrometers. Pavlenko, V. A., Ozerov, L. N., Rafal'son, A. E. *Sov. Phys. Tech. Phys.* **13**, 431(1968), trans. from *Zh. Tekh. Fiz.* **38**, 581 (1968).
Modern aspects of mass spectrometry Reed, R. I. *Book. Proc. of 2nd. NATO Adv. Study Inst. of Mass Spect. on theory and design and applications,* July 1966, Plenum Press, New York, 1968.
Time-of-flight mass spectrometer. *Instrum. Rev.* **15**, 45 (1968).
Instruments and their applications in mass spectrometry. Beynon, J. H., and Fontaine, A. E. *Instrum. Rev.* **14**, 501 (1967).
Bibliography and index on vacuum and low pressure measurement January 1960 – December 1965. Brombacher, W. G., NBS. *Monograph. No. 35 Suppl.* **1**, 1967.

2 Techniques and modifications

New time-of-flight mass spectrometer. Katzenstein, H. S., and Friedland, S. S., *Rev. Sci. Instr.* **26**, 324 (1955).
Time-of-flight mass spectrometer with improved resolution. Wiley, W. C., and McLaren., I. H. *Rev. Sci. Instr.* **26**, 1150 (1955).
Method of analysis of an ion beam. Heym, A., Joseph, C., and Loude, J. F., *Helv. Phys. Acta* **34**, 444 (1961).
Improved sampling and recording systems in gas chromatography – time-of-flight mass spectrometry. Ebert, A. A., Jr. *Anal. Chem.* **33**, 1865 (1961).
Resistance strip magnetic electron multiplier. Goodrich, G. W., and Wiley, W. C., *Rev. Sci. Instr.* **32**, 846 (1961).
Experimental methods for studying free radicals using a time-of-flight mass spectrometer. Harrington, D. B., *Fifth International Symposium on Free Radicals, Uppsala*(1961).
A modified time-of-flight mass spectrometer for studying ion-molecule or neutral particle-molecule interactions. Lehrle, R. S., Robb, J. C., and Thomas, D. W., *J. Sci. Inst.* **39**, 458 (1962).
High temperature Knudsen effusion cell assembly. Rauh, E. G. Sadler, R. C., and Thorn R. J., *Argonne Nat. Lab.,* ANL Report 6536 (1962).
Continuous ion source for a time-of-flight mass spectrometer. Studier, M. H., *Rev. Sci. Inst.* **34**, 1367 (1963).
A high temperature inlet manifold for coupling a gas chromatograph to the time-of-flight mass spectrometer. Miller, D. O., *Anal. Chem.* **35**, 2033 (1963).
Calibration of a flyable mass spectrometer for nitrogen and oxygen atoms. Cohen, H. A., Morgan, J. E., Narcissi, R. S., and Schiff, H. I., (McGill Univ., Montreal, Canada), AFCRL-63-437 (1963).
Time-of-flight mass spectrometer with sampling conversion of output signal. Anufriev,

G. S., and Mamyrin, B. A., *Prilsurg i Tekh Eksper.* 150 (1964).

Source and multiplier modifications to a time-of-flight mass spectrometer to improve sensitivity. Bitner, E. D., Rohwedder, W. K., and Selke, E., *Appl. Spectr.* **18**, 134 (1964).

Mass spectrometric studies of reactions in flames. I. Beam formation and mass dependence in sampling 1 atm gases. Greene, F. T., Brewer, J., and Milne, T. A., *J. Chem. Phys.* **40**, 1488 (1964).

Observation and indentification of ion dissociation processes occurring in the drift tube of a time-of-flight mass spectrometer. Hunt, W. W., Jr., Huffmann, R. E., and McGee, K. E., *Rev. Sci. Instr.* **35**, 82 (1964).

Time-of-flight mass spectrometer adapted for studying charge transfer, ion dissociation and photoionisation. Hunt, W. W., Jr., Huffmann, R. E., Saari, J., Wassel, G., Betts, J. F., Paufve, E. H., Wyess, W., and Fluegge, R. A., *Rev. Sci. Instr.* **35**, 88 (1964).

Increased photographic sensitivity for time-resolved mass spectrometer data recording. Meyer, R. T., *Rev. Sci. Instr.* **35**, 1064 (1964).

Pulse generator system for fast-reaction experiments using a time-of-flight mass spectrometer. Riney, J. S., and Gardiner, W. C., Jr., *Rev. Sci. Instr.* **35**, 384 (1964).

Simple display system for recording time-of-flight mass spectra. Lincoln, K. A., *Rev. Sci. Instr.* **35**, 1688 (1964).

Synchronisation of a flash lamp and mass spectrometer apparatus for fast reaction studies. Freese, J. M., and Meyer, R. T.,(Scandia Corporation, New Mexico), SC-RR-65-185, Contract AT(29-1)-789 (1965).

Magnetic tape recording of analytical data. Issenberg, P., Bazinet, M. L., and Merritt, C., Jr., *Anal. Chem.* **37**, 1074 (1965).

A time-of-flight mass spectrometer. Matus L., *Magyar Tud. Akad. Kozp. Fiz. Kut. Int. Kozlemen* **13**, 251 (1965).

Stationary film recording of time-resolved mass spectra. Moulton, D., McL., and Michael, J. V., *Rev. Sci. Instr.* **36**, 226 (1965).

Basic properties of electron multiplier ion detection and pulse counting methods in mass spectrometry. Dietz, L. A., *Rev. Sci. Instr.* **36**, 1763 (1965).

Rapid determination of electron impact ionisation and appearance potentials. Martignoni, P., Morgan, R. L., and Cason, C., *Rev. Sci. Instr.* **36**, 1783 (1965).

Time-of-flight spectrometer for laser surface interaction studies. Bernal, E., Levine, L. P., and Ready, J. F., *Rev. Sci. Instr.* **37**, 938 (1966).

Time-of-flight mass spectrometers. Jayaram, R., *Mass Spectrometry,* Plenum Press, New York, 1966, Chapter 3.

Time-of-flight mass spectrometers. Blauth, E. W., *Dynamic Mass Spectrometers,* Elsevier, Amsterdam, 1966, Chapter 4.

Technique to improve signal-to-noise ratio of electron multiplier in a mass spectrometer. Young, J. R., *Rev. Sci. Instr.* **37**, 1414 (1966).

Computational methods for resolution of mass spectra. Luenberger, D. G., and Dennis, U. E., *Anal. Chem.* **38**, 715 (1966).

Computer-assisted identification of mass spectrometric, flash filament signals. McCarroll, B., *Rev. Sci. Instr.* **38**, 444 (1967).

Shock tube time-of-flight mass spectrometer apparatus with cryosorption pumping. Ryason, P. R., *Rev. Sci. Instr.* **38**, 607 (1967).

Method of operating an ion source for a time-of-flight mass spectrometer. Studier, M. H., *US Pat.* 3,296,434 (1967).

Modification of a time-of-flight mass spectrometer for investigation of ion-molecule reactions at elevated pressures. Futrell, J. H., Tiernan, T. O., Abramson, F. P., and Miller, C. D., *Rev. Sci. Instr.* **39**, 340 (1968).

Improvement of spectral baseline stability for a time-of-flight mass spectrometer operated at elevated pressures. Miller, C. D., Tiernan, T. O., and Futrell, J. H., *Rev. Sci. Instrum.*, **40**, 503 (1969).

Multigrid energy and mass analyser for plasma diagnoses. Inoue, N., Leloup, C., and Uchida, T., *Intern. J. Mass. Spect. Ion Phys.* **2**, 85 (1969).

A time-of-flight mass spectrometer. Wadham, P. R., *Mass Spectrometry*, Butterworths, London, *Proc. Symp. Mass Spect., Enfield College of Technol., 5th and 6th July 1967* 53, N.1.4 (1968).

Mass discrimination in a time-of-flight mass spectrometer. Part 1. Geometric mass discrimination at magnetic electron multiplier. Hunt, W. W., McGee, K. E., Streeter, J. K., and Maughan, S. E., *Rev. Sci. Instrum.* **39**, 1793 (1968).

A small airborne mass spectrometer — the MKH-5401. Kudryavtsev, G. N., Levina, G. N., Martynkevich, G. M., Ozerov, L. N., Pavlenko, V. A., and Rafal'son, A. E., *Tsentral Aerolog. Obs. Trudy.* **77**, 33 (1967), (trans. from Russian) ATD Press **7**, 48 (1968).

Ion source and accelerator assembly for a time-of-flight mass spectrometer. Gohlke, R. S., and Karle, F. J. *US Pat.* 3,390,264 (Dow Chemical Co.) (1968) *Nucl. Sci. Abstr.* **22**, 4127 No. 40489 (1968).

Time-of-flight mass spectrometry apparatus having a plurality of chambers with electrically resistive coatings. Gohlke, R. S., and Karle, F. J. *US Pat.* 3,394,252 (Dow Chemical Co.) (1968), *Nucl. Sci. Abstr.* **23**, 77, No. 744 (1969).

Building a blanking generator for the Bendix multiplier M-105 G6 in a time-of-flight mass spectrometer (German). Petrick, W. H., Pfeifer, J. P., and Schindler, R. N., N68-21371 JUL-507-PC Institut Physik. Chem. Kernforschungs Anlage, Juelich, W. Germany 1967. *Nasa Star* **6**, 1630 (1968); *Nucl. Sci. Abstr.* **22**, 3169 (1968).

A time-of-flight mass spectrometer suitable for ionospheric composition investigations. Diem. H. T., N67-17331 NASA-CR-93184 SR-309 Ionosphere Lab., Penn. Univ., Univ. Park, Penn., USA. 1967, *NASA Star* **6**, 1176 (1968).

A spectrometer for studying atomic collisions in the energy range 5-100 keV. Woolley, R. L., Warner, A. G., Poole, D. H., and Hancock, R. *Conf. on Heavy Particle Collisions, Queens's Univ. of Belfast, N. Ireland 1-3 April 1968,* 279 (1968).

Blanking circuit for a magnetic electron multiplier in a time-of-flight mass spectrometer. Haumann, J. R., and Studier, M. H., *Rev. Sci. Instrum.* **39**, 169 (1968).

Synchronous operation of mass spectrometer. Kende, P., *Rev. Sci. Instrum.* **39**, 270 (1968).

3 Mass spectral data

Application of mass spectrometry to structure problems. VI. Nucleosides. Biemann, K. and McCloskey, J., *J. Am. Chem. Soc.* **84**, 2005 (1962).

Fluorine compounds of xenon and radon. Chernick, C. L., Claassen, H. H., Fields, P. R.,

Hyman, H. H., Malm, J. G., Manning, W. M. Matheson, H. S., Quarterman, L. A., Schreiner, F., Selig, H. H., Sheft, I., Siegel, S., Sloth, E. N., Stein, L., Studier, M. H., Weeks, J. L., and Zirin, M. H., *Science* **138**, 136 (1962).

Volatile hop constituents: identification of methyl deca-4-enoate and methyl deca-4,8-dienoate. Buttery, R. G., Lundin, R. E., McFadden, W. H., Jahnsen, V. J., and Kealy, M. P., *Chem. Ind.* 1981 (1963).

Further studies on the mass spectra of butanols: 3-monodeutero-1-butanol and 4-monodeutero-1-butanol. McFadden, W. H., Black, D. R., and Corse, J. W., *J. Phys. Chem.* **67**, 1517 (1963).

Volatiles from oranges. I Hydrocarbons identified by infrared, n.m.r. and mass spectra. Black, D. R., McFadden, W. H., Lundin, R. E. Schultz, T. H., and Teranishi, R., *J. Food. Sci.* **28**, 541 (1963).

Xenon tetroxide's mass spectrum. Huston, J. L., Sloth, E. N., and Studier, M. H., *Science* **143**, 1161 (1964).

Mass spectrometric identification of decomposition products of polytetrafluoroethylene and polyfluoroethylene propylene. Kupel, R. E., Nolan, M., Keenan, R. G., Hite, M., and Scheel, L. D., *Anal. Chem.* **36**, 386 (1964).

Silicon-fluorine chemistry. I. Silicon difluoride and the perfluorosilones. Timms, P. L., Kent, R. A., Ehlert, T. C., and Margrave, J. L., *J. Am. Chem. Soc.* **87**, 2824 (1965).

Silicon-fluorine chemistry. II Boron fluorides. Timms, P. L., Ehlert, T. C., Margrave, J. L., Brinckman, F. E., Farrar, T. C., and Coyle, T. D., *J. Am. Chem. Soc.* **87**, 3819 (1965).

Rearrangement of some piperidine N-oxides to hexahydro-1,2-oxazepines. Quinn, L. D., and Shelburne, F. A., *J. Org. Chem.* **30**, 3135 (1965).

Mass spectrometric studies of unsaturated methyl esters. Rohwedder, W. K., Mabrook, A. F,, and Selke, E., *J. Phys. Chem.* **69**, 1711 (1965).

Specific rearrangements in the mass spectra of neopentyl esters. McFadden, W. H., Stevens, K. L., Meyerson, S., Karabatos, G. J., and Orzech, C. E., *J. Phys. Chem.* **69**, 1742 (1965).

Organic compounds in carbonaceous chondrites. Anders, E., Hoyatsu, R., and Studier, M. H., *Science* **149**, 1455 (1965).

Correlations and anomalies in mass spectra: lactoses. McFadden, W. H., Day, E. A., and Diamond, M. J., *Anal. Chem.* **37**, 89 (1965).

Correlations and anomalies in mass spectra: thioesters. McFadden, W. H., Seifert, R. M., and Wasserman, J., *Anal. Chem.* **37**, 560 (1965).

Doubly charged transition metal carbonyl ions. Winters, R. E., and Kiser R. W., *J. Phys. Chem.* **70**, 1680 (1966).

Radiation chemistry of cyclopentane. Hughes, B. M., and Hanrahan, R. J., *J. Phys. Chem.* **69**, 2707 (1965).

Observation of astatine compounds by time-of-flight mass spectrometry. Appelman, E. H., Sloth, E. N., and Studier, M. H., *Inorg, Chem.* **5**, 766 (1966).

Feasibility of analysing ablator blow-off gases with a time-of-flight mass spectrometer. Lincoln. K. A., *NASA Star* **5**, 325 (1966).

Synthesis of thioborine. Kirk, R. W., and Timms, P. L., *Chem. Commun.* 18 (1967).

Boron-fluorine chemistry. I. BF and some derivatives. Timms, P. L., *J. Am. Chem. Soc.* **89**, 1629 (1967).

Chemistry of silicon difluoride. Margrave, J. L., and Thompson, J. C., *Science* **155**, 669 (1967).

Gaseous oxides and oxyacids of iodine and xenon : mass spectra. Studier, M. H., and Huston, J. L., *J. Phys. Chem.* **71**, 457 (1967).
Perbromic acid mass spectrum. Studier, M. H., *J. Am. Chem. Soc.* **90**, 1901 (1968)
Metastable ions in the Bendix time-of-flight. Johnsen, R. H., and Mooberry, F., TID-24192 Florida State Univ., Tallahassee, USA *Nucl. Sci. Abstr.* **22**, 846 No. 8408 (1968).
Application of mass spectrometry to forensic toxicology. Part I. Acidic and neutral drugs. De, P. K., and Unberger, C. J., *Appl. Spectr.* **22**, 353 No. 19 (1968). Abstract 7th National Meeting of the Society for Applied Spectroscopy.
Xenon trioxide difluoride. Mass spectrum. Huston, J. L., *Inorg. Nucl. Chem. Letters* **4**, 29 (1968).
Residual gas analysis and leak detection by time-of-flight measurements on neutral, metastable atoms and molecules. Crosby, D. A., and Zorn, J. C., *J. Vac. Sci. Technol.* **5**, 173 Abstr. No. 3-2 (1968). 15th Nat. Vacuum. Symp., Pittsburg, USA. 29 October – 1 November (1968).
The mass spectra of amino acid and peptide derivatives. Jones, J. H. *Quart. Rev.* **22**, 302 (1968).
Analyses of pyrimidine and purine bases by a combination of paper chromatography and time-of-flight mass spectrometry. Studier, M. H., Hayatsu, R., and Fuse, K., N68-28051 NASA-CR-95356 Chicago Univ., Ill., USA 1967, *NASA Star* **6**, 2600 (1968).
Study of the method of separation of isotopes using time-of-flight principle. Ramana, J. V., and Navalkar, M. P., *Proc. Nucl. Phys. Solid State Phys. Symp., Kanpur, India 27 February – 3 March,* 543 (1967), *Nucl. Sci. Abstr.* **22**, 3171, (1968).
Volatiles from oranges. Part 6. Constituents of the essence identified by mass spectra. Schultz, T. H., Black, D. R., Bomben, J. L., Mon. T. R., and Teranishi, R. *J. Food Sci.* **32**, 698 (1967).
Volatile from grapes. Identification of volatiles from concord essence. Stern D. J., Lee, A., McFadden, W. H., and Stevens, K. L., *J. Agr. Food Chem.* **15**, 1100 (1967).
Mass spectrographic analysis of exhaust products from an air aspirating diesel fuel burner. Wessler, M. A., *Diss. Abstr.* **27**, 3955 (1967).
Nuclear chemistry group progress report 1967. Muga, M. L., ORO-2843-12 Florida Univ. Gainesville, USA 1967. *Nucl. Sci. Abstr.* **22**, 995 (1968).

4 Gas chromatograph inlet

Time-of-flight mass spectrometry and gas-liquid partition chromatography. Gohlke, R. S., *Anal. Chem.* **31**, 535 (1959).
Some applications of mass spectrometry to lipid research. Dutton, H. J., *J. Am. Oil Chem. Soc.* **38**, 660 (1961).
Application of time-of-flight mass spectrometry and gas chromatography to reaction studies. Levy, E. J. Miller, E. D., and Beggs, W. S., *Anal. Chem.* **35**, 946 (1963).
Use of capillary gas chromatography with a time-of-flight mass spectrometer. Black, D. R., Day, J. C. McFadden, W. H., and Teranishi, R., *J. Food Sci.* **28**, 316 (1963).
Volatiles from strawberries. I. Mass spectral identification of the more volatile components. Black, D. R., Corse, J. W., McFadden, W. H., Morgan, A. I., and Teranishi, R., *J. Food Sci.* **28**, 478 (1963).
Fast-scan mass spectrometry with capillary gas-liquid chromatography in investigation of fruit volatiles. McFadden, W. H., and Teranishi, R., *Nature* **200**, 329 (1963).
Constituents of hop oil. Buttery, R. G., McFadden, W. H., Teranishi, R., Kealy, M. P., and

Mon., T. R., *Nature* **200**, 435 (1963).

A rapid method for isolation and identification of sesquiterpene hydrocarbons in cold-pressed grapefruit oil. Hunter, G. L. K., and Brogden, W. B., Jr., *Anal. Chem.* **36**, 1122 (1964).

Volatiles from oranges. II. Constituents of juice identified by mass spectra. Corse, J. W., Kilpatrick, P. W., McFadden, W. H., Schultz, T. H., and Teranishi, R., *J. Food Sci.* **29**, 790 (1964).

Gas chromatographic separation of oxygen difluoride. Yee, D. Y., *J. Gas Chromatog.* **3**, 314 (1965).

Cyclobutane compounds. III. The ionic addition of hydrogen chloride, hydrogen bromide and hydrogen iodide to allene and methylacetylene. Griesbaum, K., Naegele, W. and Wanless, G. G., *J. Am. Chem. Soc.* **87**, 3151 (1965).

Radiation chemistry of perfluorocyclohexane and perfluorocyclobutane. Fallgatter, M. B. and Hanrahan, R. J., *J. Phys. Chem.* **69**, 2059 (1965).

Identification of aldehydes and ketones from the time-of-flight mass spectra of their 2,4-dinitrophenylhydrazones. Light, J. F., and McKinney, R. W., *Appl. Spectr.* **19**, 193 (1965).

Mass spectra of isocyanates. Ruth, J. M. and Philippe, R. J., *Anal. Chem.* **38**, 720 (1966).

Use of activated charcoal to trap gas chromatographic fractions for mass spectrometry and to introduce volatile compounds into a mass spectrometer. Damico, J. M., Sphon, J. A., and Wong, N. P., *Anal. Chem.* **39**, 1045 (1967).

Analysis of proteins, peptides and amino acids by pyrolysis – gas chromatography – time-of-flight mass spectrometry. Merritt, C., and Robertson, D. H., *J. Gas Chromatog.* **5**, 96 (1967).

Multichannel open tabular columns. Merritt, C., Jr., and Walsh, J. T., *J. Gas Chromatog.* **5**, 420 (1967).

Introduction of gas chromatographic samples to a mass spectrometer. McFadden, W. H., *Separation Sci.* **1**, 723 (1966).

Analyses of complex mixtures of hydrocarbons by time-of-flight mass spectromery/open tube chromatography. Studier, M. H., and Hayatsu, R. *Anal. Chem.* **40**, 1011 (1968).

Analysis of oxidised paraffins by combined techniques. Brown, R. A., Kay, M. I., Kelliher, J. M. and Dietz, W. A., *Anal. Chem.* **39**, 1805 (1967).

5 Shock tubes

Shock wave studies by mass spectrometry. I. Thermal decomposition of nitrous oxide Bradley, J. N., and Kistiakowsky, G. B., *J. Chem. Phys.* **35**, 256 (1961).

Shock wave studies by mass spectrometry. II. Polymerisation and oxidation of acetylene. Bradley, J. N., and Kistiakowsky, G. B., *J. Chem. Phys.* **35**, 264 (1961).

Decomposition of nitromethane in a shock tube. Bradley, J. N., *Trans. Faraday Soc.* **57**, 1750 (1961).

Mass spectral studies of kinetics behind shock waves. I. Thermal dissociation of chlorine Diesen, R. W., and Felmlee, W. J., *J. Chem. Phys.* **39**, 2115 (1963).

Mass spectral studies of kinetics behind shock waves. II. Thermal decomposition of hydrazine. Diesen, R. W., *J. Chem. Phys.* **39**, 2121 (1963).

Observation and kinetic investigation of the NF and NF_2 radicals. Diesen, R. W., *J. Chem. Phys.* **41**, 3256 (1964).

Mechanism of chemi-ionisation in hydrocarbon oxidations. Kistiakowsky, G. B., and Michael, J. V., *J. Chem. Phys.* **40**, 1447 (1964).

Shock wave studies by mass spectrometry. III. Description of apparatus; data on the oxidation of acetylene and of methane. Dove, J. E., and Moulton, D. McL., *Proc. Roy. Soc. (London) A,* **283**, 216 (1965).

Kinetics of the nitrous oxide decomposition by mass spectrometry. A study to evaluate gas-sampling methods behind reflected shock waves. Modica, A. P., *J. Phys. Chem.* **69**, 2111 (1965).

Mechanism of the acetylene-oxygen reaction in shock waves. Glass, G. P., Kistiakowsky, G. B., Michael, J. V., and Niki, H., *J. Chem. Phys.* **42**, 608 (1965).

Thermal decomposition of acetylene in shock waves. Gay, I. D., Kistiakowsky, G. B., Michael, J. V., and Niki, H., *J. Chem. Phys.* **43**, 1720 (1965).

Pyrolysis and oxidation of formaldehyde in shock waves. Gay, I. D., Glass, G. P., Kistiakowsky, G. B., and Niki, H., *J. Chem. Phys.* **43**, 4017 (1965).

Decomposition and oxidation of C_2F_4 behind shock waves. Modica, A. P., and Lagraffe, J. E., *J. Chem. Phys.* **43**, 3383 (1965).

Comment on NF_2 decomposition behind shock waves. Diesen, R. W., *J. Chem. Phys.* **45**, 759 (1966).

Pyrolysis of ethylene in shock waves. Gay, I. D., Kern, R. D., Kistiakowsky, G. B., and Niki, H., *J. Chem. Phys.* **45**, 2371 (1966).

Mass spectrometric and ultraviolet absorption study of CHF_3 decomposition behind shock waves. Modica, A. P., and Lagraffe, J. E., *J. Chem. Phys.* **44**, 3375 (1966).

Mass spectrometry studies behind shock waves. III. The thermal dissociation of fluorine. Diesen, R. W., *J. Chem. Phys.* **44**, 3662 (1966).

Mass spectrometric observation of ions formed during shock wave heating of gaseous Kr and Xe. Cresswell, R., DiValentin, M., and Dove, J. R., *Phys. Fluids* **9**, 2285 (1966).

Ethylene-oxygen reaction in shock waves. Gay, I. D., Glass, G. P., Kern, R. D., and Kistiakowsky, G. B., *J. Chem. Phys.* **47**, 313 (1967).

Kinetics of the difluoromethylene-NO reaction. I. Modica, A. P., *J. Chem. Phys.* **46**, 3663 (1967).

Kinetics of the reaction of fluorine with difluoroamino radicals and the dissociation of fluorine. Diesen, R. W., *J. Phys. Chem.* **72**, 108 (1968).

Mass spectral studies of kinetics behind shock waves direct sampling and flash photolysis of NO_2. Felmlee, W. J., Petrella, R. V., and Diesen, R. W., AD-482393 DOW-SL-175311A 1966. *US Govt. Res. Dev. Rep.* **68**, 79 (1968).

Time-resolved mass spectrometry. Thermal and photo decomposition of NO_2. Diesen, R. W., Petrella, R. V., and Felmlee, W. J., American Chem. Soc. National meeting, San Francisco, April, 1968 Sect. U No. 53 (1968).

6   Flash photolysis

A mass spectrometric study of flash photochemical reactions. Kistiakowsky, G. B., and Kydd, P. H., *J. Am. Chem. Soc.* **79**, 4825 (1957).

Benzyne. Berry, R. S., Clardy, J., and Schafer, M. E., *J. Am. Chem. Soc.* **86**, 2738 (1964).

The dimerisation of gaseous benzyne. Schafer, M. E., and Berry, R. S., *J. Am. Chem. Soc.* **87**, 4497 (1965).

1,4-Dehydrobenzene. A stable species. Berry, R. S., Clardy, J., and Schafer, M. E., *Tetrahedron Letters* 1003 (1965).

Decomposition of benzenediazonium-3-carbonylate: transient 1,3-dehydrobenzene. Berry, R. S., Clardy, J., and Schafer, M. E., *Tetrahedron Letters* 1011 (1965).

Dimerisation of gaseous benzyne. Schafer, M. E., *Diss. Abstr.* 27, 444B (1966).

Flash photolysis and time-resolved mass spectrometry. I. Detection of the hydroxyl radical. Meyer, R. T., *J. Chem. Phys.* 46, 967 (1967).

Reactions of excited iodine atoms. I. ($^2P_{1/2}$) in the flash photolysis of methyl iodide. Meyer, R. T., *J. Chem. Phys.* 46, 4146 (1967).

Flash photolysis and time-resolved mass spectrometry. Part 2. Decomposition of methyl iodide and reactivity of $I(^2P_{1/2})$ atoms. Meyer R. T., *J. Phys. Chem.* 72, 1583 (1968).

Comments of flash photolysis and time-resolved mass spectrometry. Meyer, R. T., SC-M-68-125 Sandia Corp., Albuquerque, N. Mex., USA 24P 1968. Nucl. Sci. Abstr. 22, 4384 (1968).

Flash-photolysed reactions monitored by time-resolved mass spectrometry. Meyer, R. T., American Chem. Soc. National meeting, San Francisco, April, 1968. Sect. U No. 55 (1968).

7 Flash pyrolysis and laser evaporation

Products of the flash pyrolysis of phenol-formaldehyde by time-of-flight mass spectrometry. Friedman, H., *J. Appl. Polymer Sci.* 9, 651 (1965).

Flash-vaporisation of solid materials for mass spectrometry by intense thermal radiation. Lincoln, K. A., *Anal. Chem.* 37, 541 (1965).

Flash pyrolysis of solid-fuel materials by thermal radiation. Lincoln, K. A., *Pyrodynamics* 2, 133 (1965).

Mass spectrometer using a laser probe. Knox, B. E., and Vastola, F. J., *Chem. Eng. News* 44, 48 (1966).

Laser microprobe allied to mass spectrometry. Knox, B. E., and Vastola, F. J., *Laser Focus* 3, 15 (1967).

Mass spectrometric analysis of carbon species generated by laser evaporation. Zavitsanos, P. D., General Electric, Missile and Space Division, USA, Tech. Inf. Series R67SD11 (1967).

Laser heating of coal particles in the source of a time-of-flight mass spectrometer. Joy, W. K., Ladner, W. R., and Pritchard, E., *Nature* 217, 640 (1968).

Laser ionisation. Vastola, F. J., *Eastern analytical symposium 13-15 November, 1968, Am. Chem. Soc., Soc. Appl. Spectrosc. and Am., Microchem. Soc.*, New York, USA. Abstracts (1968), p. 33.

Analysis of Surface by laser evaporation. Vastola, F. J., *Appl. Spectr.* 22, 374 (1968).

Instrumentation for time-resolved mass spectrometry application to laser vaporisation of solid materials. Lincoln, K. A., N68-30322 USNRDL-TR-68-36 AD-669453 Naval Radiological Defence Lab., San Francisco, California, USA 1968, *NASA Star* 6, 3084 (1968). *US Govt. Res. Dev. Rep.* 68, 97 (1968).

Improved instrumentation for time-resolved mass spectrometry with application to laser-vaporisation of solid materials. Lincoln, K. A., *Intern. J. Mass Spect. Ion Phys.* 2, 75 (1969).

Ionisation of organic solids by laser irradiation. Vastola, F. J., and Pirone, A. J., *Advan. Mass Spect. Proc. Conf. Berlin 1967.* 4, 107, (1968) Inst. of Petroleum London and Elsevier, Amsterdam.

Mass spectrometric analysis of carbon species generated by laser evaporation. Zavitsanos,

P. D., *Carbon* **6**, 731 (1968).

Mass spectrometric studies of laser-induced vaporisation. Part 1. Selenium. Knox, B. E., *Mat. Res. Bull.* **3**, 329 (1968).

An instrument for mass analysis using a laser. Fenner, N. C., and Daly, N. R., *J. Mater Sci.* **3**, 259 (1968).

A laser used for mass analysis. Fenner, N. C., and Daly, N. R., *Lasers Unconventional Optics J.* **8**, 87 (1967).

8 Flame and molecular beam studies

Mass spectrometric detection of polymers in supersonic molecular beams. Greene, F. T., and Milne, T. A., *J. Chem. Phys.* **39**, 3150 (1963).

Composition profile of the diethyl ether/air two-stage reaction stabilised in a flat-flame burner. Agnew, J. T., and Agnew, W. G., *10. Intern. Sym. Combustion, Cambridge, England, 1964,* published 1965, 123 (*Chem. Abstr.* **63**, 16204f).

Mass spectrometric studies of reactions in flames. II. Quantitative sampling of free radicals from one-atmosphere flames. Milne, T. A., and Greene, F. T., *J. Chem. Phys.* **44**, 2444 (1966).

Identification of flame ions by time-of-flight mass spectrometry. King, I. R., and Scheurich, J. T., *Rev. Sci. Instr.* **37**, 1219 (1966).

Mass spectrometric observations of argon clusters in nozzle beams. I. General behaviour and equilibrium dimer concentrations. Greene, F. T., and Milne, T. A., *US Govt. Res. Rept.*, AD-637-177, **41**, N19, 74 (1966).

Mass spectrometric detection of dimers of nitric oxide and other polyatomic molecules. Milne, T. A., and Greene, F. T., *J. Chem. Phys.* **47**, 3668 (1967).

Mass spectrometric observations of argon clusters in nozzle beams. Part I. General behaviour and equilibrium dimer concentrations. Milne, T. A., and Greene, F. T., *J. Chem. Phys.* **47**, 4095 (1967).

The oxidation of sodium, potassium, and caesium in flames. Carabetta, R., and Kaskan, W. E., *J. Phys. Chem.* **72**, 2483 (1968).

Mass spectrometer study of metal-containing flames. Milne, T. A., and Greene, F. T., AD-664832 Midwest Res. Inst., Kansas City, Mo., USA 1968. US Govt. *Res. Dev. Rep.* **68**, 149 (1968).

9 Miscellaneous reaction kinetic studies

Photochemical oxidations. I. Ethyl iodide. Heicklen, J., and Johnston, H. S., *J. Am. Chem. Soc.* **84**, 4394 (1962).

Mass spectrometric investigation of the high temperature reaction between nickel and chlorine. McKinley, J. D., *J. Chem. Phys.* **40**, 120 (1964).

Mass spectrometric investigation of the nickel-bromine surface reaction. McKinley, J. D., *J. Chem. Phys.* **40**, 576 (1964).

Mass spectrometric investigation of the yttrium chlorine surface reaction. McKinley, J. D., *J. Chem. Phys.* **41**, 2814 (1964).

Mass spectral study of the production of methylamine from azo-methane. Wacks. M. E., *J. Phys. Chem.* **68**, 2725 (1964).

Mass spectrometric study of desorption of oxygen from tungsten. Chuikov, B. A., Ptushinc'kii, and Yu., G., *Ukr. Fiz. Zh.* **9**, 1035 (1964) (*Chem. Abstr.* **62**, 52f (1965)).

On the role of the nitroxyl molecule in the reaction of hydrogen atoms with nitric oxide.

Kohout, F. C., and Lampe, F. W., *J. Am. Chem. Soc.* **87**, 5795 (1965).

Dynamic response measurements using a time-of-flight mass spectrometer. Douglas, J. M., *Chem. Eng. Sci.* **20**, 1142 (1965).

Mass spectrometric investigation of reactions of oxygen atoms with hydrogen and ammonia. Potter, E., and Wong, E. L., Lewis Research Centre, Cleveland, Ohio. OTS Document, NASA TND-2648 (1965).

Silicon-fluorine chemistry. IV. The reaction of silicon difluoride with aromatic compounds. Timms. P. L., Stump, D. D., Kent, R. A., and Margrave, J. L., *J. Am. Chem. Soc.* **88**, 940 (1966).

Mass spectrometric investigation of the nickel-fluorine surface reaction. McKinley, J. D., *J. Chem. Phys.* **45**, 1690 (1966).

Reactions of oxygen (^3P) atoms with formaldehyde. Niki, H., *J. Chem. Phys.* **45**, 2330 (1966).

Reaction of oxygen (^3P) atoms with diacetylene. Niki, H., and Weinstock, B., *J. Chem. Phys.* **45**, 3468 (1966).

Application of adiabatic compression apparatus for study of thermally induced homogeneous gas phase reactions. Barnes, V. M., Jr., Henderson, U. V., and Rhodes, H. A., *Rev. Sci. Instr.* **37**, 294 (1966).

Kinetic study on the reaction of oxygen atoms with dimethyl ether by time-of-flight mass spectrometry. Mori, S., Oishi, K., and Takeyaki, Y., *Bull. Inst. Chem. Res. Kyoto Univ.* **4**, 341 (1966).

Application of a time-of-flight mass spectrometer in studies of arc-quenching reactions in sulphur hexafluoride gas. Kamatani, A., and Miyamoto, T., *Elect. Eng. Japan* **84**, 30 (1966).

Mass spectroscopic investigation of the reaction of oxygen atoms with methane. Wong, E. L., and Potter, A. E., *Can. J. Chem.* **45**, 367 (1967).

A mass spectrometer investigation of the nitroxyl (HNO) molecule. Kohout, F. C., *Diss. Abstr.* **27**, 3889-B (1967).

Metastable transitions observed in a time-of-flight mass spectrometer. Dugger, D. L., and Kiser, R. W., *J. Chem. Phys.* **47**, 5054 (1967).

Some reactions of silicon dichloride. Timms, P. L., *Inorg. Chem.* **7**, 387 (1968).

Langmuir vaporisation studies by time-resolved mass spectrometry. Meyer, R. T., and Ames, L. L., *Am. Chem. Soc. National meeting, San Francisco, April, 1968* Sect. S No. 266 (1968).

Determination of the rate constant of fast reaction from the transient concentration profile by means of a time-of-flight mass spectrometer. Takezaki, Y., and Mori, S. *Int. Conf. Photochemistry, Munich, 6-9 September 1967, Ber. Bunsenges. Phys. Chem.* **72**, 157 (1968).

Mass spectrometric investigation of discharge flow system $N_2 + CH_4$. Biordi, J. C., RNL-2310-214 Rad. Res. Labs., Mellon Inst., Pittsburg, Pa., USA. 1967.

Determination of the rate constant of fast reaction from the transient concentration profile by means of a time-of-flight mass spectrometer. Takezaki, Y., and Mori, S., N68-26082 Inst. for Chemical Research Kyoto Univ., Japan. Max. Planck. Inst. Kohlenforschung Photochem. Part 1, 73 (1967). *NASA Star* **6**, 2409 (1968).

Application of a time-of-flight mass spectrometer in studying the kinetics of fast processes in high temperature decomposition of ammonium perchlorate. Korobeynichev, O. P., Boldyrev, V. V., and Karpenko, Yu. Yu., *Fiz. Goreniya, Vzryva*, **1**, 33 (1968), ATD

Press, *Abstr. Library of Congress,* Washington, DC, USA **7**, 43 (1968).

Determination of micro-reactor volume from the transient concentration profile by means of a time-of-flight mass spectrometer and the application to the rate constant measurement of fast reactions, $O + NO_2 = NO + O_2$ and $N + NO = N_2 + O$. Takezaki, Y., and Mori, S., *Bull. Inst. Chem. Res.* (Japan) **45**, 388 (1968).

Formation of hydrogen in the photolysis of diborane at 1849Å. Bufalini, M., and Todd, J. E., *J. Phys. Chem.* **72**, 3367 (1968).

Application of time-resolved mass spectrometry to problems in high temperature chemistry. Meyer, R. T., and Ames, L. L., *Advan. Chem. Ser.* **72**, Am. Chem. Soc., Washington, DC, USA 1968, p. 301.

Reaction of ketene with atomic hydrogen and oxygen. Carr, R. W., Gay, I. D., Glass, G. P., and Niki, H. *J. Chem. Phys.* **49**, 846 (1968).

Reaction of atomic oxygen with methyl radicals. Niki, H., Daby, E. E., and Weinstock, B. *J. Chem. Phys.* **48**, 5729 (1968).

Mass spectrometric study of the reaction of atomic hydrogen with acetaldehyde. McKnight, C., Niki, H., and Weinstock, B. *J. Chem. Phys.* **47**, 5219 (1967).

Reaction of oxygen atoms with chlorine. Niki, H., and Weinstock, B., *J. Chem. Phys.* **47**, 3249 (1967).

10 Thermal analysis

Mass spectrometric thermal analysis (MTA), Langer, H.,G., and Gohlke, R. S., *Anal. Chem.,* **35**, 1301 (1963).

Obtaining the mass spectra of non-volatile or thermally unstable compounds. Gohlke, R. S., *Chem. Ind.* 946 (1963).

Mass spectrometric techniques applied to the investigation of the thermal degradation of polymers. Wacks, M. E., *Conf. High Temp. Polymers Fluid Research,* ASD-TDR-62-372 (1962), 201.

Mass spectrometric studies of the thermal decomposition of poly (carbon monofluoride). Kuriakose, A. K., and Margrave, J. L., *Inorg. Chem.* **4**, 1639 (1965).

Thermal decomposition by mass spectrometry. Collin, J. E., and Delplace, A., *Bull. Soc. Chim. Belges* **75**, 281 and 304 (1966).

Thermal analysis by mass spectrometry. Gohlke, R. S., and Langer, H. G., *Fortschr. Chem. Fortsch.* **6**, 515 (1966).

Thermogravimetric-mass spectrometric analysis. Zitomer, F. *Anal. Chem.* **40**, 1091 (1968).

The particles resulting from polytetrafluoroethylene (PTFE) pyrolysis in air. Coleman, W. E., Scheel, L. D., and Gorski, C. H., *Amer. Ind. Hygiene. Assoc. J.* **29**, 54 (1968).

Time-of-flight analysis of caesium ions in caesium. Popescu, I., Niculescu, N., and Popescu, A. *Rev. Roum. Phys.* **13**, 51 (1968). *Sci. Abstr. Ser. A Phys. Abstr.* **71**, 3371 (1968).

Thermal degradation of proteins studied by mass spectrometry. Kasarda, D. D., and Black, D. R., *Biopolymers* **6**, 1001 (1968).

Mass spectrometric analysis of the pyrolysis products of polymeric materials. Goldstein, H. W., UCRL-13332 UC-4-4 General Electric Co., Philadelphia, Pa., USA, 1967.

Application of mass spectrometry to the investigation of vaporisation (Roumanian). Topor, D., *Stud. Cercet. Chim.* **16**, 3 (1968).

Mass spectrometric study of the thermal decomposition of ammonium perchlorate. Maycock, J. N., PaiVerneker,V. R., and Jacobs, P. W. M., *J. Chem. Phys.* **46**, 2857 (1967).

Mass spectrometric study of the thermal decomposition of ammonium perchlorate. Pai Verneker, V. R., and Maycock, J. N., *J. Chem. Phys.* **47**, 3618 (1967).

The thermal decomposition of nitronium perchlorate. Maycock, J. N., and Pai Verneker, V. R., *J. Phys. Chem.* **71**, 4077 (1967).

The effects of interfaces on the thermal degradation of polymer-metal composites. Schmidt, G. A., and Gaulin, C. A., *J. Appl. Polymer Sci.* **11**, 357 (1967).

11 Knudsen cell

Uranium monosulphide. II. Mass spectrometric study of its vaporisation. Cater, E. D., Rauh, E. G., and Thorn, R. J., *J. Chem. Phys.* **35**, 619 (1961).

Vapour pressures of scandium, yttrium and lanthanum. Ackermann, R. J., and Rauh, E. G., *J. Chem. Phys.* **36**, 448 (1962).

Effect of oxygen on the vapour pressure of uranium. Ackermann, R. J., Rauh, E. G., and Thorn, R. J., *J. Chem. Phys.* **37**, 2693 (1962).

Application of time-of-flight mass spectrometry to the study of inorganic materials at elevated temperatures. Goldstein, H. W., Sommer, A., Walsh, P. N., and White, D. O., *Advan. Mass Spec.* **2**, 110 (1963).

Thermodynamic properties of actinide elements. Ackermann, R. J., and Thorn R. J., *Proc. Symp. Intern. Atomic Energy Agency, Vienna, 1962,* published 1963, p. 39.

Vaporisation studies of thorium, uranium and plutonium metals and oxides. Ackermann, R. J., and Thorn, R. J., *Proc. Symp. Intern. Atomic Energy Agency, Vienna, 1962,* published 1963, p. 445.

Vaporisation of the rare earth oxides at elevated temperatures. Ames, L. L., Goldstein, H. W., Walsh, P. N. and White D., *Proc. Symp. Intern. Atomic Energy Agency, Vienna, 1962,* published 1963 p. 417.

Vaporisation studies on uranium-sulphur and uranium-oxygen-sulphur systems. Cater, E. D., Rauh, E. G., and Thorn, R. J., *Proc. Symp. Intern. Atomic Energy Agency, Vienna, 1962,* published 1963, p. 487.

A thermodynamic study of the thorium-oxygen system at high temperatures. Ackermann, R. J., Rauh, E. G., Thorn, R. J. and Cannon, M. C., *J. Phys Chem.* **67**, 762 (1963).

A thermodynamic study of the tungsten-oxygen system at high temperatures. Ackermann, R. J., and Rauh, E. G., *J. Phys. Chem.* **67**, 2596 (1963).

A literature review of mass spectrometer-thermochemical technique applicable to the analysis of vapour species over solid inorganic materials. Redman, J. D. (Oak Ridge Nat. Lab.), Report ORNL-TM-989 (1964).

Thermodynamic properties of gaseous yttrium monoxide. Correlation of bonding in Group III transition-metal monoxides. Ackermann, R. J., Rauh, E. G., and Thorn, R. J., *J. Chem. Phys.* **40**, 883 (1964).

Dissociation energies of diatomic molecules of the transition elements. I. Nickel. Kant, A., *J. Chem. Phys.* **41**, 1872 (1964).

The use of a pulsed time-of-flight mass spectrometer for investigating the evaporation of compounds. Mostovskii, A. A., and Sakseev, D. A., *Zhurnal Tekhn. Fiziki* **34**, 1321 (1964).

Dissociation energies of diatomic molecules of the transition elements. II. Titanium, chromium, manganese and cobalt. Kant, A., and Strauss, B., *J. Chem. Phys.* **41**, 3806 (1964).

Vaporisation thermodynamics and dissociation energy of lanthanum monosulphide. Cater,

E. D., Lee, T., Johnson, E., Rauh, E. G., and Eick, H., *J. Phys. Chem.* **69**, 2684 (1965).

Mass spectrum of yttrium chloride vapour. McKinley, J. D., *J. Chem. Phys.* **42**, 2245 (1965).

Comments on pressure of uranium over uranium/uranium dioxide system. Ackermann, R. J., Rauh, E. G., and Thorn, R. J., *J. Chem. Phys.* **42**, 2630 (1965).

Time-of-flight mass spectrometric determination of dissociation energy of BeF. Hildebrand, D. L., and Murad, E., *J. Chem. Phys.* **44**, 1524 (1966).

High temperature equilibrium in germanium-nickel systems. Dissociation energy of GeNi. Kant. A., *J. Chem. Phys.* **44**, 2451 (1966).

Atomisation energies of the polymers of germanium. Kant, A., and Strauss, B. H., *J. Chem. Phys.* **45**, 822 (1966).

Dissociation energy of Cr_2. Kant, A., and Strauss, B., *J. Chem. Phys.* **45**, 3161 (1966).

Mass spectrometric studies at high temperature. IX. Sublimation pressure of copper (II) fluoride. Kent, R. A., McDonald, J. D., and Margrave, J. L., *J. Phys. Chem.* **70**, 874 (1966).

Vaporisation, thermodynamics, and dissociation energy of lanthanum monosulphide. Part 2. Cater, E. D., and Steiger, R. P., *J. Phys. Chem.* **72**, 2231 (1968).

Thermochemistry of UOs. Evaporations of $US-UO_2$ mixtures. On the attainment of equilibrium in Knudsen cells. Cater, E. H., Rauh, E. G., and Thorn, R. J., *J. Chem. Phys.* **49**, 4650 (1968).

Mass spectrometric studies of gaseous systems Au-Ni, Au-Co and Au-Fe, and dissociation energies of AuNi, AuCo, and AuFe. Kant, A., *J. Chem. Phys.* **49**, 5144 (1968).

Mass spectrometric studies of volatile Al, Sc, Co, Y, Zr and rare earth chelates. McDonald, J. D., and Margrave, J. L., *J. Less-Common Metals* **14**, 236 (1968).

Mass spectrometric studies at high temperatures. Part 25. Vapour composition over $LaCl_3$, $EuCl_3$ and $LuCl_3$ and stabilities of the trichloride dimers. Hastie, J. W., Ficalora, P., and Margrave, J. L., *J. Less-Common Metals* **14**, 83 (1968).

Vaporisation, thermodynamics, and dissociation energy of yttrium monosulphide. Steiger, R. A;. *Diss. Abstr.* **28**, 3246-B (1968).

Mass spectrometric study of tetrakis (trifluorophosphine) nickel (O). Krassoi, M. A., Kiser, R. W., and Clarke, R. J. *J. Am. Chem. Soc.* **89**, 3653 (1967).

12 Measurement of appearance and ionisation potentials

Electron impact spectroscopy of ethylene sulphide and ethylenimine. Gallegos, E., and Kiser, R., *J. Phys. Chem.* **65**, 1177 (1961).

Studies of the shapes of ionisation efficiency curves of multiply charged monatomic ions. I. Instrumentation and relative electronic transition probabilities for krypton and xenon ions. Kiser, R. W., *J. Chem. Phys.* **36**, 2964 (1962).

Electron impact spectroscopy of the four and five-membered saturated heterocyclic compounds containing nitrogen, oxygen and sulphur. Gallegos, E., and Kiser, R. W., *J. Phys. Chem.* **66**, 136 (1962).

Electron impact spectroscopy of tetramethylgermanium, trimethylsilane and dimethylmercury. Hobrock, B., and Kiser, R. W., *J. Phys. Chem.* **66**, 155 (1962).

A technique for the rapid determination of ionisation and appearance potentials. Kiser, R. W., and Gallegos, E. J., *J. Phys. Chem.* **66**, 947 (1962).

Ionisation potentials of cyclopropyl radical and cyclopropyl cyanide (communications to the Editor). Kiser, R. W., and Gallegos, E. J. *J. Phys. Chem.* **66**, 957 (1962).

The ionisation potential of hydrogen disulphide (H_2S_2). Kiser, R. W., and Gallegos, E. J., *J. Phys. Chem.* **66**, 1214 (1962).

Electron impact spectroscopy of propylene sulphide. Hobrock, B., and Kiser, R. W., *J. Phys. Chem.* **66**, 1551 (1962).

Electron impact spectroscopy of sulphur compounds: 1,2-thiabutane, 2-thiapentane and 2,3-dithiabutane. Hobrock, B., and Kiser, R. W., *J. Phys. Chem.* **66**, 1648 (1962).

Electron impact spectroscopy of some substituted oxiranes. Wada, Y., and Kiser, R. W., *J. Phys. Chem.* **66**, 1652 (1962).

The ionisation potential of (iso)thiocyanic acid. Shenkel, R., Hobrock, B., and Kiser, R. W., *J. Phys. Chem.* **66**, 2074 (1962).

Ionisation and dissociation processes in phosphorus trichloride and diphosphorus tetrachloride. Sandoval, A., Moser, H., and Kiser, R. W., *J. Phys. Chem.* **67**, 124 (1963).

Electron impact investigation of sulphur compounds. II. 3-methyl-2-thiabutane, 4-thia-1-pentane and 3,4-dithiahexane. Hobrock, B., and Kiser, R. W., *J. Phys. Chem.* **67**, 648 (1963).

Electron impact investigations of sulphur compounds. III. 2-thiapropane and 2,3,4-trithiapentane. Hobrock, B. and Kiser, R. W., *J. Phys. Chem.* **67**, 1283 (1964).

Singly and doubly charged ions from methyl and ethyl isothiocyanates by electron impact. Hobrock, B., Shenkel, R., and Kiser, R. W., *J. Phys. Chem.* **67**, 1684, (1963).

Comparative ionisation by MeV heavy ions and keV electrons. Schuler, R. H., and Stuber, F. A., *J. Chem. Phys.* **40**, 2035 (1964) (erratum: *J. Chem. Phys.* **41**, 901 (1964)).

Appearance potentials of positive and negative ions by mass spectrometry. Melton, C. E., and Hamill, W. H., *J. Chem. Phys.* **41**, 546 (1964).

Ionisation-efficiency curves for molecular and fragment ions for methane and the methyl halides. Tsuda, S., Melton, C. E., and Hamill, W. H., *J. Chem. Phys.* **41**, 689 (1964).

Appearance potentials by the retarding potential difference method for secondary ions produced by excited-neutral, excited ion-neutral, and ion-neutral reactions. Hamill, W. H., and Melton, C. E., *J. Chem. Phys.* **41**, 1469 (1964).

Structure of ionisation-efficiency curves near threshold from alkanes and alkyl halides. Tsuda, S. and Hamill, W. H., *J. Chem. Phys.* **41**, 2713 (1964).

A study of threshold behaviour of electron impact induced ion and dissociation ionisation of simple gaseous molecules. Glick, R. A., and Llewellyn, J. A., Bulletin No. 14, Division of Biology and Medicine, Inst. of Molecular Biophysics, USAEC (1964), Doc. No. FSU-2690-14.

Structure in the ionisation efficiency curves of Ar_2^+ by pulsed mass spectrometry. Becker, P. M., and Lampe, F. W., *J. Am. Chem. Soc.* **86**, 5347 (1964).

Electron impact studies of some trihalomethanes. Hobrock, D., and Kiser, R. W., *J. Phys. Chem.* **68**, 575 (1964).

Mass spectral study of trimethyl borate. Wada, Y., and Kiser, R. W., *J. Phys. Chem.* **68**, 1588 (1964).

A mass spectral study of some alkyl-substituted phosphines. Wada, Y., and Kiser, R. W., *J. Phys Chem.* **68**, 2290 (1964).

Stabilization energies of substituted methyl cations – the effect of strong demand on the resonance order. Lampe, F. W., Martin, R. H., and Taft, R. W., *J. Am. Chem. Soc.* **87**, 2490 (1965).

An electron impact study of ionisation and dissociation of trimethylsilanes. Hess, G. G., Lampe, F. W., and Sommer, L. H., *J. Am. Chem. Soc.* **87**, 5327 (1965).

Ionisation potential and heat of formation of the difluoromethylene radical. Pottie, R. F., *J. Chem. Phys.* **42**, 2607 (1965).

Multiple ionisation in neon, argon, krypton and xenon. Stuber, F. A., *J. Chem. Phys.* **42**, 2639 (1965).

Ions produced by electron impact with the dimetallic carbonyls of cobalt and manganese. Winters, R. E., and Kiser, R. W., *J. Phys. Chem.* **69**, 1618 (1965).

Ionisation potentials and mass spectra of cyclopentadienyl-molybdenum dicarbonyl nitrosyl and 1,3-cyclohexadieneiron tricarbonyl. Winters, R. E., and Kiser, R. W., *J. Phys. Chem.* **69**, 3198 (1965).

Ionisation and dissociation of ruthenium and osmium tetroxides. Dillard, J. G., and Kiser, R. W., *J. Phys. Chem.* **69**, 3893 (1965).

Mass spectrometric investigations of the synthesis stability and energetics of the low temperature oxygen fluorides. Malone, T. J. and McGee, H. A., Jr., *J. Phys. Chem.* **69**, 4338 (1965).

An electron impact study of ionisation and dissociation in methoxy- and halogen-substituted methanes. Martin, R. H., Lampe, F. W., and Taft, R. W., *J. Am. Chem. Soc.* **88**, 1353 (1966).

An electron impact investigation of some alkyl phosphate esters. Bafus, D. A., Gallegos, E. J., and Kiser, R. W., *J. Phys. Chem.* **70**, 2614 (1966).

Analysis of ionisation efficiency curves and multiple ionisation processes. Shadoff, L. A., *Diss. Abs.* **27**, 1445B (1966).

Mass spectrometric studies of the synthesis, energetics and cryogenic stability of the lower boron hydrides. Wilson, J. H., and McGee, H. A., *J. Chem. Phys.* **46**, 1444 (1967).

Electron gun performance using a modified appearance potential technique. Sarzone, G., *NASA Star* **N20**, 3982 (1966). (*Mass Spec. Bull.* **1**, 781 (1967)).

Ionisation potentials of free radicals formed by electron impact. Methylstannyl radicals. Lampe, F. W., and Niehaus, A. *J. Chem. Phys.* **49**, 2949 (1968).

Appearance potentials, ionisation potentials and heats of formation for perfluorosilanes and perfluoroborosilanes. McDonald, J. D., Williams, C. H., Thompson, J. C., and Margrave, J. L., *Advan. Chem. Ser. 72,* Am. Chem. Soc., Washington DC, USA (1968). p. 261. *Symp.,152nd Mtg. A.C.S. New York 15-16 September 1966.*

Appearance potentials and kinetic energies of ions from N_2, CO, and NO. Hierl, P. M., and Franklin, J. L., *J. Chem. Phys.* **47**, 3154 (1967).

Measurement of the translational energy of ions with a time-of-flight mass spectrometer. Franklin, J. L., Hierl, P. M., and Whan, D. A., *J. Chem. Phys.* **47**, 3148 (1967).

Investigations on positive ion emission from tungsten by time-of-flight mass spectrometer. Kaposi, O., Riedel, M., and Baktai, Gy., *Proc. 14th Colloquium Spectroscopicum Internationale 7-12 Aug. 1967, Debrecen, Hungary,* Vol. 3., Hilger, London, 1967, p. 1541.

Investigation of the positive ion emission of tungsten filament with a time-of-flight mass spectrometer. Part 2.(Hungarian). Kaposi, O., Riedel, M., and Matus, L., *Magy. Chem. Foly.* **74**, 429 (1968).

13   Ion-molecule collision studies

Charge transfer studies with a time-of-flight mass spectrometer: cross-section for N_2^+ on N_2 at 2650 eV. Hunt, W. W., Novock, R., and Stinnett, A. J., Tech. Rept. AFCRL-63-709 (1963).

Application of time-of-flight mass spectrometry to the examination of ion-molecule interactions. Homer, J. B., Lehrle, R. S., Robb, J. C., Takahasi, M., and Thomas, D. W., *Advan. Mass Spectrom.* **2**, 503 (1963).

Specific reaction rate of the second-order formation of Ar_2^+. Lampe, F. W., and Hess, G. G., *J. Am. Chem. Soc.* **86**, 2952 (1964).

Ion-molecule reactions studied by time-of-flight mass spectrometry. Hand, C. W., and von Weyssenhoff, H., *Can. J. Chem.* **42**, 195 (1964).

Ion-molecule reactions studied by time-of-flight mass spectrometry. II. Reactions in CO-D_2 and CH_4-D_2 mixtures. Hand, C. W., and von Weyssenhoff, H., *Can. J. Chem.* **42**, 2385 (1964).

Ion dissociation in the drift tube of a time-of-flight mass spectrometer: spurious fragments arising from charge transfer and dissociation reactions of retarded ions. Hunt, W. W., Jr., and McGee, K. E., *J. Chem. Phys.* **41**, 2709 (1964).

Gas-phase ion-molecule interactions involving atom transfer; limitations of the orbiting theory in accounting for the variation of cross-section with energy. Homer, J. H., Lehrle, R. S., Robb, J. C., and Thomas, D. W., *Nature* **202**, 795 (1964).

Observation of the products of ionic collision processes and ion decomposition in a linear, pulsed time-of-flight mass spectrometer. Ferguson, R. E., McCulloh, K. E., and Rosenstock, H. M., *J. Chem. Phys.* **42**, 100 (1965).

Mass spectrometric study of the bimolecular formation of diatomic argon ion. Becker, P. M., and Lampe, F. W., *J. Chem. Phys.* **42**, 3857 (1965).

Pulsed mass spectrometric study of the bimolecular formation of N_3^+. Cress, M. C., Becker, P. M., and Lampe, F. W., *J. Chem. Phys.* **44**, 2212 (1966).

Mass spectrometric study of ion-molecule reactions of ethanol and methanol. Sieck, L. W., Abramson, F. P., and Futrell, J. H., *J. Chem. Phys.* **45**, 2859 (1966).

High transmissions and dual electron beam ion sources for mass spectrometry. Melton, C. E., *J. Sci. Instr.* **43**, 927 (1966).

Study of ion-molecule reactions at high pressures using a very sensitive T.O.F. mass spectrometer. Conway, D. C., *Am. Chem. Soc. 156th Nat. meeting, Atlantic City, New Jersey, USA 8-13 September 1968.* Abstr. No. Phys. 42, 1968.

Mass spectrometric investigation of ion-molecule reactions in cyclohexane. Abramson, F. P., and Futrell, J. H., *J. Phys. Chem.* **71**, 3791 (1967).

Low-field drift velocities and reactions of nitrogen ions in nitrogen. McKnight, L. G., McAfee, K. B., and Sipler, D. P., *Phys. Rev.* **164**, 62 (1968).

Positive ion emission from polymers under ion impact. Dillon, A. F., Lehrle, R. S., Robb, J. C., and Thomas, D. W., *Advan. Mass Spectr. Proc. Conf. Berlin 1967* **4**, 477 (1968), Inst. of Petroleum, London and Elsevier, Amsterdam.

The mechanism of dissociative charge transfer reactions. Lehrle, R. S., Robb, J. C., Scarborough, J. and Thomas, D. W., *Advan. Mass Spectr. Proc. Conf. Berlin 1967.* **4**, 687 (1968), Inst. of Petroleum, London and Elsevier, Amsterdam.

Measurement of the cross-section for the reactions $H^- + H_2O = OH^- + H_2$ and $D^- + D_2O = OD^- + D_2$ at incident ion energies near 2 eV. Stockdale, J. A. D., Compton, R. N., and Reinhardt, P. W., *Phys. Rev. Letters* **21**, 664 (1968).

Development of a time-of-flight mass spectrometer working at 10^{-8} torr and its use for the study of a charge transfer reaction (French). Zvenigorosky, A., *Thesis,* University of Toulouse, France, 1967; *Vacuum Index, Index, Biblio du Vide* **3**, 102 (1968).

The krypton-radiosensitised reaction of deuterium atoms with ethylene. Tewarson, A., and Lampe, F. W., *J. Phys. Chem.* **72**, 3261 (1968).

Chemical reaction kinematics. Part 8. Cross-sections of some D-atom transfer reactions in the energy range 1-100 eV. Hyatt, D., and Lacmann, K. *Z. Naturforsch* **23A**, 2019 (1968).

Ion-molecule reactions in alpha-particle-irradiated methane and water vapour. Fluegge, R. A., *J. Chem. Phys.* **50**, 4373 (1969).

14 Negative ion studies

Autodetachment of electrons in sulphur hexafluoride. Edelson, D., Griffiths, J. E., and McAfee, K. B., Jr., *J. Chem. Phys.* **37**, 917 (1962).

Non-dissociative electron capture in complex molecules and negative ion lifetimes. Compton, R. N., Christophorou, L. G., Hurst, G. S., and Reinhart, P. W., *J. Chem. Phys.* **45**, 4634 (1966).

Secondary electron capture by hexafluoroacetone. Thynne, J. C. J., *Chem. Commun.* 1075 (1968).

Ion-molecule reactions of negative ions. Part 1. Negative ions of sulphur. Dillard, J. C., and Franklyn, J. L. *J. Chem. Phys.* **48**, 2349 (1968).

The formation of negative ions in allyl alcohol and acrolein (French). Bouby, L., Compton, R. N., and Souleyrol, A., *Compt. Rend. Ser. C.* **266**, 1250 (1968).

Ionisation and dissociation of hexafluoroethane, and of 1,1,1-trifluoroethane and fluoroform, by electron impact. MacNeil, K. A. G. and Thynne, J. C. J., *Intern. J. Mass Spectrom. Ion Phys.* **2**, 1 (1969).

15 Miscellaneous

Measurement of breathing by analysis of respiratory gases by time-of-flight mass spectrometry. Caldwell, P. R. B., and Graves, J. B. (Aerospace Med. Div. Wright Patterson Airforce Base, Ohio). US Govt. Reprt. (1965), AD 608339 (AMRL-TR-64-84).

Observations on rats exposed to a space cabin atmosphere for two weeks. Felic, P. *Aerospace Medicine* **36**, 858 (1965).

Direct observation of the decomposition of multiply charged ions into singly charged fragments. McCulloh, K. E., Sharp, T. E., and Rosenstock, H. M., *J. Chem. Phys.* **42**, 3501 (1965).

Comparative stabilities of gaseous alane, gallane and indane. Breisacher, P., and Siegel, B., *J. Am. Chem. Soc.* **87**, 4255 (1965).

Basic studies in radiation technology. Final report, JLI-2901-75, USAEC, Johnston Labs. Inc., Baltimore (1966).

Low energy investigation of radiation chemistry. Kiser, R. W., *Mass Spectrom. Bull.* **1**, 346 (1967).

Classification and dynamic calibration of ionisation sensors and detectors. Alcalay, J. A., and Knuth, F. L., *4th Int. Vacuum Congress Manchester, 17-20 April 1968* (Abstr.) Session 38 *Vacuum* **18**, 119 (1968).

Motion of an ion swarm in a cylindrically symmetric drift tube mass spectrometer. Moseley, J. T., and Gatland, I. R., AD-671230 SQUID-TR-GIT-3-PU Project Squid Headquarters, Purdue, Univ., Lafayette, Ind., USA, 1968. *US Govt. Res. Dev. Rep.* **68**, 66 (1968).

Index

Absorption spectroscopy 38
Acetaldehyde 236
Acetylene 49-54, 193, 231, 232
Acrolein 242
Actinides 237
Actinometry 44
Alane 242
Aldehydes 231
Allene 231
Alloy 97, 98, 147
Alloy, cobalt-platinum 156, 159, 160
Allyl alcohol 242
Aluminium 238
Amino acids 230, 231
Ammonia 29, 190, 235
Ammonia, deutero 87-104
Ammonium nitrate 185, 189, 190, 194, 196
Ammonium perchlorate 235, 236, 237
Amplitude limiting device for recorders 215-217
Analogue scanners ... 89, 106, 164, 165, 167, 173, 202, 205, 208, 211, 212, 215
Appearance potential 4, 80, 84, 97, 105-137, 227, 238-240
Argon .. 32, 33, 44, 47, 49, 50, 78, 79, 90, 95, 97, 142, 143, 144, 234, 240
Astatine compounds 229
Azomethane 234
Benzene 165-172, 196
Benzenediazonium-3-carbonxlate 233
Benzonitrile 62
Benzyne 232, 233
Berylium fluoride 238
Blanking generator 228
Bond dissociation energy 37, 118-137, 237, 238
Boron 229
Boron hydrides 240
Bromine 234
Butanols 229
Caesium 234, 236
Calcium carbonate 17, 21
Calibration 226
Camera, drum 41, 184, 185
Camera, polaroid 41
Carbonaceous chondrite 229
Carbon dioxide 17, 19, 20, 21, 190, 195
Carbon disulphide 118
Carbon monoxide 21, 126, 137, 195, 240, 241
Carbonyl fluoride 106, 118, 119, 128
Carbonyl selenide 38

Carbonyl sulphide 118
Catalyst 90, 92, 98, 102
Catalytic chemistry 87
Charge transfer reactions 116-137, 227, 241
Chemionisation 232
Chlorine 231, 234, 236
Chloroform 171
Chromiun 237, 238
Clusters 234
Coal 15, 18-21, 183-197, 233
Cobalt 155, 156, 159, 160, 237, 238
Cobalt carbonyl 240
Combustion 19-21
Computer 70, 227
Copper 155, 159, 185, 189, 194
Copper (II) fluoride 238
Cross-linking agents 175-181
Crucible heater for direct inlet probe 180
Cyclobutane compounds 231
Cyclohexane 241
1,3-Cyclohexadieneiron tricarbonyl 240
Cyclopentane 229
Cyclopentadienyl-molybdenum dicarbonyl nitrosyl 240
Cyclopropyl cyanide 238
Data recording 41, 42
Deconvolution 105-137, 142, 145
Dehydrobenzene 232, 233
Deuterium 95, 100, 102-4, 241, 242
Diacetylene 193, 235
Diborane 236
Diethyl ether 234
Difluoromethylene 232
Digitised output 203
Diglycidyl ether of bis-phenol A 175, 176
Dihydroxydiphenylmethane 175
Dimethyl ether 235
Dimethylmercury 238
Dinitrophenylhydrazones 231
Dioxane 170, 171
Diphosphorus tetrachloride 239
Dissociative energy transfer 97
Dissociative resonance capture 105, 106, 118, 134
Dithiabutane 239
Dithiahexane 239
Electrical discharge 29
Electromagnetic interference 44
Electron affinity 131
Electron capture detector 163, 170

Electron gun 75, 76
Electronics 26, 71-77, 205, 211-214, 215-217
Electron multiplier.......... 65, 150, 152, 204, 206, 208,
                                                         211, 226-8
Electron optics 71
Electron radiolysis 29
Energy focussing 200
Energy transfer 143
Epichlorhydrin 175
Epoxy resins 175-181
Ethane 18, 48-54
Ethanol 241
Ethylbenzene 165-171
Ethylene 49-57, 232, 241
Ethylene sulphide 238
Ethylenimine 238
Europium trichloride.. 238
Excited species............................. 33, 52, 53, 82, 126
Fast reaction 227
Ferric oxide... 90
Field-evaporation 148-150, 152
Field-ionisation 147, 162
Field-ion microscope 147-162
Filament, current supply 75
Filament, iron catalyst 89
Filament, rhenium 75
Flames 227, 234
Flame ionisation detector 165-171
Flash desorption 44
Flash heating... 44
Flash lamp.. 42-44, 56, 227
Flash photolysis 26, 37-57, 232, 233
Flight-time 139-145, 153
Fluorine 84, 119, 228, 229, 232, 235
Fluorocarbons 203
Fluoroform 232, 242
Forensic toxicology.. 230
Formaldehyde 232, 235
Free radicals.................. 26, 29, 38, 194, 196, 226, 234
    H-atom 29-31, 50, 196, 234
    OH 39, 233
    NH_2 29-31
    NH, ND 99, 101
    NF, NF_2 .... 231, 232
    CH_3 .... 37, 47-57, 193, 195, 196, 236
    C_2H_5 55
    CF_3 126
    CF_2 239
    $Pb(CH_3)_x$ 38, 48-57
    $CH_2Pb(CH_3)_3$ 48, 54, 55
    Cyclopropyl.. 238

Fluoroformyl 118, 121, 128
    Methyl stannyl 240
    Trifluoroacetyl... 126
Fuel 15, 233
Furnace, electron bombardment 2,3
Gallane 242
Gas chromatography .. 37, 44, 163-174, 203, 205, 206,
                                                      219-224, 230, 231
Gas chromatography, selective detector 163-174
Gas chromatography — T.O.F. mass spectrometer
   inlet 164, 165, 219-224, 226, 230, 231
Gas-solid reactions... 15-21
Germanium 87, 238
Gold 155-157, 180, 238
Graphite..1-13
Helium 29, 31-33
n-Heptane 171-173
Hexafluoroacetone..................106, 121-130, 136, 242
Hexafluoroethane 242
Hexane 206, 220-222
Hop oil... 230
Hydrazine 231
Hydrocarbons 26, 48-53, 90, 185, 189-191, 195,
                                                           229, 231, 232
Hydrogen 39, 49-54, 60, 62, 90, 91, 98, 142-144,
                                              157, 159, 235, 236, 241
Hydrogen bromide 231
Hydrogen chloride..197, 231
Hydrogen disulphide 239
Hydrogen iodide 231
Indane... 242
Inlet system, Bendix 843A 175, 180, 208
Iodine 230
Ion bombardment...90
Ion-electron secondary emission ratio 204, 207
Ion energy spread...199, 200
Ionisation cross-section 4, 24, 29, 30, 35
Ionisation potential 23, 29, 97, 98, 105-137, 192,
                                                          227, 238-240
Ionisation threshold 105-137
Ion-molecule reactions .. 30, 35, 70, 80, 118, 139, 226,
                                                         228, 240-242
Ion optics... 71
Ion probe technique...87-104
Ion-pair formation 105
Ions per cycle 39, 44
Ion spectroscopy...139-145
Ion source ... 24, 26, 28, 29, 35, 39, 56, 62, 71, 75, 88,
                                  106, 147, 184, 199, 202, 205, 226-228
Ion transmission 205, 206
Ion velocity profile 142
Iridium 155-160
Iron................................87-104, 147, 155, 160, 238

Isocyanates 231, 239
Isothiocyanic acid 239
Ketene 39, 53, 236
Ketones 231
Kinetic measurments 15-21
Knudsen cell 1-13, 226, 237-238
Krypton 44, 91, 232, 238, 240, 241
Lactoses 229
Lanthanum 237
Lanthanum monosulphide 237, 238
Lanthanum trichloride 238
Laser 15, 183-197, 227, 233, 234
Lead iodide 200
Lead tetramethyl 37-57
Leak detection 230
Linear-to-logarithmic compressor circuit 211-214
Lipids 230
Lutetium trichloride 238
Magnetic deflection mass spectrometer 38, 199-209
Magnetic tape recording 227
Manganese 189, 237
Marganese carbonyl 240
Mass discrimination 228
Mass spectral data 228-230
Metastables 26, 31-33, 68-70, 235
Metastable ions 23, 62, 65, 230
Methane 48-57, 194, 195, 232, 235, 239, 241, 242
Methanol 241
Methylacetylene 231
Methylamine 234
2-Methylanthracene 187, 194, 196
Methyl deca-4-enoate 229
Methyl deca-4,8-dienoate 229
Methyl esters 229
Methyl iodide 39, 233
Methyl isobutyl ketone 172
Methyl n-propyl ketone 171
3-Methyl-2-thiabutane 239
Microanalysis 161
Microbalance target technique 2, 11
Molecular beans 234
Molecular separator 219
Molybdenum 155-160
Momentum transfer 143
Monoenegetic electrons 23
Multiplier discrimination 207, 208
Nadic maleic anhydride 176, 178
Negative ions 90, 105-137, 239, 242
Neon 44, 47, 78, 147, 156, 159, 240
Neopentyl esters 229
Neutral fragments 31-35

Nickel 155, 159, 234, 235, 238
Nitrobenzene 137
Nitric oxide 190, 232, 234, 236, 240
Nitrogen .. 88, 90, 95, 98, 99, 103, 190, 226, 235, 240, 241
Nitrogen dioxide 39, 44, 190, 232, 236
Nitromethane 171, 231
Nitronium perchlorate 237
Nitrous oxide 190, 231, 232
Nitroxyl 234, 235
Nucleosides 228
Oscilloscope 41, 42, 152, 155, 167, 184, 185, 205
Oscilloscope photography 41, 152, 185, 227
Osmium tetroxide 240
Oxiranes 239
Oxygen 39, 226, 232-237
Oxygen fluorides 231, 240
Paper chromatography 230
Paraffins 195, 231
Peak intensification 41
Peak width 106, 107, 110
Pentafluorosulphur chloride 106, 132-136
Peptides 230, 231
Perbromic acid 230
Perfluoroborosilanes 240
Perfluorocyclobutane 231
Perfluorocyclohexane 231
Perfluoroethylene 232
Perfluorokerosene 200
Perfluoromethylcyclohexane 137
Perfluorosilanes 240
Petrol 172
Phenol-formaldehyde 233
Phosphines 239
Phosphorus 160
Phosphorous trichloride 239
Photoelectrons 67
Photoelectron spectrometry 71-85
Photoionisation 59-70, 227
Photoionisation, light source 60
Photoions 67
Photolysis 37
Piperidine N-oxides 229
Platinum 87, 155-160, 180
Plutonium 237
Polyacetylenes 194
Poly (carbon monofluoride) 236
Polyethylene ('Polythene') 185, 186, 195
Polyethylene glycol 185, 188, 196
Polyfluoroethylene propylene 229
Polymers 87, 97, 99, 175, 183, 194, 196, 234, 236, 237, 241

Polystyrene 188, 195, 196
Polytetrafluoroethylene ('Teflon') 185, 186, 191, 195, 229, 236
Potassium 234
Propanol, iso- 171
Propylacetate 171
Propylene sulphide 238
Propenyl 193
Proteins 231, 236
Pulse radiolysis 23-35
Purine 230
Pyrene 186, 187, 194, 196
Pyrimidine 230
Pyrolysis 15, 16, 59, 195, 232, 236
Pyrolysis, flash 233, 234
Quadrupole mass spectrometer 38, 199-209
Radiation chemistry 231, 242
Radon 228
Rare earth oxides 237
Rare earth chelates 238
Raster circuit 42
Reaction cross-section 97, 124
Reaction kinetic studies 234-236
Reaction vessel 42-44
Residual gas analysis 230
Resolution 199-209
Resonance attachment 105
Respiratory gases 242
R.P.D. measurements 71-85, 105, 117, 137
Ruthenium tetroxide 240
Sampling 227, 232
Scandium 237, 238
Selenium 234
Sensitivity 41-47, 69, 75, 170, 199-209, 227
Sesquiterpene hydrocarbons 231
Shock tube 37, 45, 227, 231, 232
Signal-to-noise ratio 105, 227
Silicon 229, 235
Silicon dichloride 235
Silicon difluoride 235
Sodium 234
Space cabin atmosphere 242
Statistical fluctuations 39, 44, 139, 142
Steel 147
Styrene 196
Subtraction technique 45
Sucrose 188, 194
Sugar 185
Sulphur 237, 242
Sulphur dioxide 106, 116, 137, 190
Sulphur hexafluoride...71-85, 106, 109, 115, 123, 124, 130-132, 137, 235, 242

Surface reactions 87-104, 147
Synchronous operation 227, 228
Techniques 226-228
Temperature determination 3, 15, 175, 194
Tetrachloroethylene 137
Tetrakis (trifluorophosphine) nickel 238
Tetramethyl germanium 238
Tetramethylsilane 57
Tetranitromethane 137
Thermal analysis 236, 237
Thermal conductivity detector 220
Thermal desorption 90
Thermal ionisation 192
Thermionic detector 170
Thermodynamic data 1-13, 136, 237, 238
1,2-Thiabutane 239
Thiapentanes 239
Thiapropane 239
Thioborane 229
Thorium 237
Time-lag focussing 39, 200-202
Time-of-flight mass spectrometer
    Bendix Model 12 164
         Model 14 .... 2, 16, 26, 39, 59, 139, 211
         Models 3012, 3015 .... 106, 184, 220
Titanium 155, 156, 237
Toluene 67, 165-170, 172, 220-222
Total ion current 163
Total output integrator 220, 223
Transient species 29
Transition metal carbonyl ions 229
Trap current 73, 75, 204
Tricarbonylcyclobutadienyliron 38
Trichlorofluoroethane 171
Trifluoroacetone 137
Trifluoroethane 242
Trihalomethanes 239
Trimethyl borate 239
Trimethyl pentane 172, 173
Trimethyl phosphate 68
Trimethyl silane 238, 239
Trithiapentane 239
Tungsten 147, 154-156, 234, 237, 240
Uranium 237, 238
Uranium dioxide 238
Uranium monosulphide 237
Vanadium carbide 147
Vaporisation 1-13, 236, 238
Volatiles from grapes 230
Volatiles from oranges 229, 230, 231
Volatiles from strawberries 230
Water 90, 104, 190, 241, 242

Xenon 44, 49, 56, 228, 230, 232, 238, 240
Xenon tetroxide 229
Xenon trioxide difluoride 230
Xylene 172
Yttrium 234, 237, 238
Yttrium chloride 238
Yttrium monoxide 237
Yttrium monosulphide 238
Zinc 189, 190
Zirconium 238